Applied Fisheries

Q. J. Shammi
S. Bhatnagar
Department of Zoology
Govt. N.M.V. College, Hoshangabad, M.P.

AGROBIOS (INDIA)

Published by:
Agrobios (India)
Agro House, Behind Nasrani Cinema
Chopasani Road, Jodhpur 342 003
Phone: 91-0291-2643993, 642319
E. mail: agrobiosindia@gmail.com

AGROBIOS (INDIA)

Title: Applied Fisheries

9 788177 541144
Rs. 2500.00 / US$ .00
Published by:
AGROBIOS (INDIA)

ISBN No.: 978-81-7754-114-4

Published by: Updesh Purohit for Agrobios (India), Jodhpur
Laser typeset at: Yashee Computers, Jodhpur
Cover Design by: Reena
Printed at: Agrobios Digitals Jodhpur

Preface

Applied fisheries is now becoming an increasingly popular branch of science because of the growing demand of fish as proteinous food. Applied fisheries has been introduced in almost all the Indian Universities at under graduate and post graduate level. The present book covers all these syllabi. The present book has the following distinct features.

1. A very wide range of syllabi of the different universities have been covered in entirely, including topics which are not covered by other books.
2. A very wide range of research papers and books have been referred to and consequently some new emerging topics. *E.g.* Transgenic fishes, exotic and transplanted fishes have also been included.
3. The descriptions are very well supported by diagrams and the quantitative information in tabular form.

The author's express their sincere thanks to Dr. L.L. Sharma, Principal Govt. Narmada Mahavidyalaya, Hoshangabad, Dr. R.M. Tiwari, Head of Zoology Department, Colleagues and friends whose inspiration and encouragement helped us a great ideal.

Special thanks to my publisher Agrobios (India), especially Dr. S.S. Purohit for taking pains in bringing out the book.

We are looking forward for your valuable suggestions for the improvement of the book.

Hoshangabad

Q. J. Shammi
Mrs. S. Bhatnagar

Contents

Chapter 1

Introduction to Fish

INTRODUCTION

Fishes are cold-blooded vertebrates, which breathe by means of pharyngeal gills, propelling and balancing themselves by means of fins. They are more completely adapted to aquatic life than any other craniate. The streamlined body is an adaptation to facilitate their movement through water. They possess vertebral column, skull and distinct jaws.

They inhabit different kinds of environment ranging from deep water of the ocean to the boundless surface of the open sea and from high tropics to the receding Polar Regions. They live in fast running torrential steam, the muddy waters of the bays, brackish, estuaries, stagnant pools and in the waters several feed under the ground in caves.

The surviving fish species are most numerous of all the vertebrates, of these 41.2% inhabit the fresh water and 58.2% the marine water of the world (Table 1.1)

The whole fish study includes morphology, anatomy physiology, identification, economic importance, taxonomy, embryology, ecology and endocrinology. The term Ichthyology is derived from two Latin words "*Ichthys*" meaning fish and "logos" a discourse.

TABLE 1.1:

S. No.	Habit	Percentage
1.	Fresh water species.	41.2%
2.	Diadromous fish species.	0.6%
3.	Marine, shore and continental shelf to 200m. in warm water.	39.6%
4.	Marine, shore and continental shelf to 200m. in cold water	5.6%
5.	Deep sea, benthic below 200m	6.4%
6.	Epipelagic above 200m.	1.3%
7.	Deep sea pelagic below 200m.	5.0%

Ichthyology is the branch of science dealing with the study of fishes, with the study of commercial aspects of fishes, including the study of *hydrobiology* (Chemistry of water), *limnology*, behavior of environment, population dynamics, (Craft) use of nets and gears, methods of fish catching and detection, study of pollution, *aquaculture*, and breeding etc.

Fishes excel as a source of commerce and recreation. They form a rich source of food, and are cultured both, intensively an extensively to give the best values of their raising. In addition, they yield a number of byproducts of high economic value.

TYPICAL FISH

It is rather difficult to describe a typical fish because of their numerous groups and variations, as these vertebrates have adapted themselves to a wide range of environment. More than half of all species of vertebrates are fishes. It can be said that where is water there is a fish.

The size and structure show a great diversity due to their adaptation. They may be as small as half an inch or less (Philippine gobies) or They may be as large as 50 to 70 feet (Sharks) Shapes are also variable with an elongated snake like body (Eel) to flattened body (Skates & rays). The ocean *sunfish* has a body as deep as it is long "globula puffer", both eyes on the same side (flat fish), have horse like body (sea horse). Wide distribution of fishes have also resulted into production of variety of different kind of adaptive modification in order to meet with the stresses of the physical, chemical and the biotic environment.

HISTORY OF FISHES

The history of *Ichthyology* coincides with that of Zoology in general. Aristotle (384-322 BC) known as the father of Natural History. He had an accurate knowledge of the general structure of fishes and he distinguished them from the other aquatic *amphibians* and *mammals*. His information on the habits of fish was surprisingly accurate. He adapted the nomenclature of local fisherman to designate species of fishes but his knowledge of Ichthyology was limited. The number of fishes known to Aristotle is about 115 all of which were native of the Aegean Sea adjacent to Greece.

The knowledge of fishes in India is fairly old. The use of fishes as food is evidenced from the fish engraving and fish remains obtained from the excavations at Mohan Jodaro and Harappa of the Indus Valley (2500 to 1500 BC). The son of King Vikramaditya VI has recorded common sport fish in his book "Manasallasa" (1127 AD). He divided them into marine and fresh water forms. After Aristotle, no proper work on fishes was available for 1800 years which was a period of regression in the science of Ichthyology. G. Rondelet (1507-57) had extensive knowledge but his work was entirely limited to Mediterranean forms comprising of 197 marine and 47 fresh water fishers.

The contribution of Belon (1517-1575 AD) was based on his original observations of the fishes of Mediterranean Sea in Europe. H. Salviani (1514-1572) has given almost the same description as that of Belon.

In the next century noteworthy scholar of Ichthyology Piso (1661-1678 AD), had arranged various fish species in proper systematic manner and gave them proper scientific names. F. Willughby (1635-1672) and J. Ray in *Historia Piscium* wrote about fishes of Great Britain and of the continent especially those of Germany. Peter Artedi (1705-1735 AD) called the father of Ichthyology, was a contemporary to famous taxonomist and biologist named Carlvon, Linnaeus. The main contribution of Aretdi was a study of inter-relationships between various groups of fishes and systematic classification and recognized 47 genera and several species. He died at an early age. His unfinished work was accomplished by Linnaeus in his book (1789 AD)

Linnaeus (1707-1778) first reported on fishes in system. The first writer on Indian fishes was Bloch whose work was published in 1785 as *Naturgeschichte der Ausländischen fische* Schneider (1801) expanded this work by including several marine forms.

Bloch's Ichthyology remained for many years the standard work and proved to be the most useful guide for the students. Block was surpassed by his Liacepede. Liacepede had to contend with great difficulties in the preparation of *Histoires poison Von Humboldt* was the first European naturalist, who published papers on fishes of Ecuador, Venezuela and Mexico, Patrick Russel (1803) described 200 species of fishes of Vizagapatam.

F. Hamilton gave an account of fishes found in the river Ganges and its branches which contain 269 species of fishes. Ferreira, a Portuguese, collected specimens from Brazil which were later on named by Cuvier and his associates.

Nattever discovered the South American lung fish Lepidosiren, but Fitzinger received credit for its naming because he published a brief preliminary notice before Nattever's memoir appeared.

Louis Agassiz, a Swiss naturalist conceived a lifelong interest in Brazilian fishes. He described the structure of the different scales. Agassiz had an extensive knowledge of existing and fossil fishes and opened a new field of research of fossil forms.

DP. W. O. Ayres published papers on exotic fish Polypterus from West Africa and also on deep sea fishes. David Storer (1804-1891) described some fishes of United States. After his death this work has carried on by his son, Horatio Robinson Storar (1830-1922), especially of eastern North America Charles Girard (1822-1845) a French naturalist, began his studies on American fishes with Louis Agassiz.

His first contribution (1850-1852) dealt with the fresh water corticales, Albert Gunther (1830-1914) published papers on the fishes of British Museum in 1859-1870, covering the fishes of the entire world. David Starr Jordan (1896-1900) published an intensive description of fishes of North and Middle America, with chief emphasis on those of the United States. He was a dominant figure of America's Ichthyology.

Jordan and Gilbert conducted the basic survey of the fishes of the pacific coast in 1880. Later Day's" Fishes of India" (1875-1878) and fauna of British India, Burma and Ceylon 1889 are the significant work of that time. Jordan for his Chief contribution was the pioneer in the development of fishery biology and in the intensive study of the biology of certain commercially important fishes. Carl. H. Eigenmann (1863-1927) an American Scientist, worked on fresh water fish fauna of India, pacific coast and South America and was regarded as an eminent worker in the field of systemic zoology by Jordan.

Torteton H. Bean (1846-1916) was a major figure of Ichthyology and wrote books on fishes of Pennsylvania (1893). He worked on deep sea fishes of Atlantic and other oceans and fresh water fishes of N. American, Alaska etc.

Jordan and Evermann produced their popular *treatise* on American food and game fishes in 1902. Jordan and coworkers contributed much of what we know regarding the fishes of Mexico.

The critical work on the fresh water fishes of the country was largely confined to the basin by basin survey of the western states largely conducted by Synder (1869-1958) and other also contributed to it.

Boulenger (1904) gave a systematic account of Teleostei under the series of *Cambridge Natural History* Weber and Beaufort (1911-1915) described the fishes of Indo-Australian archipelago. In 18th and 19th centuries the science of Ichthyology received due recognition almost all over the world.

Several books were published on the fishes of individual states or regions. Notable amongst them were the fishes of North Carolina (1907) by Hugh M. Smith, T. H. Bean's treatise on the fishes of New York, (1901-1905) the fishes of the Rocky Mountain Region (1908) by Thodore D. Alison Cockeral and others.

History of Indian Fishes

The history of Indian Fishes is quite old. The use of fishes as food is evidenced from the fish engravings and fish remains obtained from the excavation of the Indus Valley.

In the 20th Century, besides Chaudhary's (1912) account of some new species of fresh water fishes of Northern India, the contribution by Hora and others (1920-56) are highly commendable, will be perpetuated in the history of Indian Ichthyology. Fish systematics have been elaborated by them. The works of Nair (1953), Haig (1950) Silas (1951-1958), Menon M. A. S. (1952) and Prasad (1962), have been a further step in this direction. Mishra's (1949, 1953, and 1962) gave a valuable contribution on fishes, with special reference to Clupeids. Menon (1963) made studies on the fishes of the Indian Museum and then gave a revised account of the fishes of the genus Garra in 1964 and reported several new forms. Jayaram (1954) revised the genus *Mystus scopolia.*

Shaw and Shtebbeare (1937) studies fishes of North Bengal. The fishes of Rajasthan were studied by Mathus (1952) and Dutta Gupta *et al.* (1961). The fishes of Meerut were studied by Sinha and Shiromani (1953); The fishes of Delhi by Majumdar (1952) and the fishes of Nainital by Choudhary and Khandelwal (1960). These works have thrown new light on the bionomics and distribution of Indian Fishes. The work on other aspects of fishes by Panniker (1958), Khanna (1970) and Jhingaran (1974) and other are notable contributions in the field of fish and fisheries.

GENERAL STRUCTURE OF FISH

1. Fishes are cold blooded vertebrates and their body is suitable for aquatic mode of life.
2. They have stream lined body with few exception.
3. They have a combination of median and paired fins.
4. There are median dorsal, anal and caudal fins.
5. There is a pair of pectoral fin and a pair of pelvic fin located posteriorly.
6. The body is covered by dermal scales, denticles or bony plates.
7. In the most primitive jawed fishes, there is autostylic jaw suspension. In the primitive sharks it was amphistylic and in the later sharks and bony fishes the jaw suspension is of hyostylic type.
8. Gills are the chief respiratory organs but sometimes accessory respiratory organs may be present.
9. Gill slits may be covered by operculum as in Teleosts. The maximum number of gill are 7 pairs.
10. In the bony fishes, the spiracle is much suppressed or it disappears completely.
11. It has one auricle and one ventricle. The venous blood is pumped to the gills for aeration.
12. There are 10 pairs of cranial nerves.
13. Their lateral line system and receptors are well developed.
14. The adult kidney is *mesonephric.*
15. Paired nostrils are present, which normally do not open into the

pharynx except in lung and lobefin fishes.

16. There is no middle ear.
17. Paired visceral arches are present. The first pair forms the upper and lower jaws, the second forms suspensorium and the rest supports the gills.
18. They are widely distributed. (Fig. 1.1)

IDENTIFICATION OF FISH

Fish identification is a process of matching the diagnostic features with the specimens of a species, which is named thereafter. Identification of a fish is based on three following types of characters.

Morphometric Character

The term suggests that it is the external measurement of the fish body. Various parts of the body are measured in the form of body length ratio. Since the older descriptions of fishes are based on rations, it is necessary to work out ratios in our material so that they can be easily compared. The various measurement of the fish body to be used for this purpose in a fish is shown in Fig. 1.1 and explained below.

The ratios are worked out by simple division of one measurement by the other. For example in a specimen with 100mm. total length, the length of head is 20mm; the ratio of the later in relation to the total length comes to five times in total length (*i.e.* 100-20). Similar method can be used for calculating other body ratios. In case, one wants to express the ratio on percentage basis it can be worked out as follows

Percentage of length of head in total length =

$$\frac{\text{Length of head}}{\text{Total length of body.}} \times 100$$

In fish an identification following terms and abbreviations are considered to describe the various parts.

1. ***Total length:*** Measured from the tip of the snout to the tip of the caudal fin. In case the lobes of the caudal fin differ in length, it is measured upto the tip of the longer lobe.
2. ***Standard length:*** Measured from the tip of the snout to the base of the caudal fin or the posterior edge of the hypural plate.
3. ***Height or depth of body:*** Measured as a straight line vertical distance from the belly to the dorsum just in front of dorsal fin or at its deepest part.
4. ***Girth length:*** It is the distance covering the circumference of the body at its deepest level.
5. ***Length of head:*** Measured from the tip of the snout of the posterior most bony extremity of the operculum, but in case of

percoids, opercular spines are also included.

6. ***Greatest height of head:*** Taken as a straight line distance between a point on the occiput and the isthmus at right angles to the longitudinal axis of head.
7. ***Greatest width of head:*** Measured as a distance through the head taken at the extreme posterior limit of the head between the opercula on either side.
8. ***Length of snout:*** Measured from the tip of the snout to the anterior border of the orbit.
9. ***Upper jaw length:*** It is the greatest length of the upper jaw.
10. ***Lower jaw length:*** It is the greatest length of the lower jaw (Fig. 1.2)

    The length measurements of the following parts may also be required, the terminology of which is as follows:

    a) *Nape*: It is the back of the neck.
    b) *Isthmus*: It is the inter space which separates the gill openings of the left and right side and which lies anterior to the breast.
    c) *Breast:* It is the region which lies posterior to the isthmus and extends up to the base of the pectoral fin. Heart lies in this region.
    d) *Occiput:* It is the point on the mid dorsal line which joins the head with the trunk.
    e) *Occipital:* It is the distance between the two perpendicular lines touching the posterior end of the orbit and the occiput.
    f) *Diameter of eye:* Measurement taken between the anterior and posterior walls of the bony orbit.
    g) *Inter orbital width or space:* Measured as the distance between the upper edges of the orbits.
    h) *Post orbital length:* It is the distance in a straight line between the posterior margin of the orbit and the posterior most edge of the opercular bone.
11. ***Pre pectoral length:*** It is a distance between the anterior most end of the body and the front point of the pectoral fin base.
12. ***Pre Anal length:*** It is the distance between the anterior most end of the body and the front point of the anal fin base.
13. ***Length of caudal peduncle:*** Measured from the last ray or end of anal fin to the base of the caudal fin.
14. ***Height of caudal peduncle:*** Measured along the vertical line at its narrowest part.
15. ***Predorsal length:*** Measured in a straight line between the tip of the snout and the insertion of first ray of the dorsal fin. (Fig 1.2)

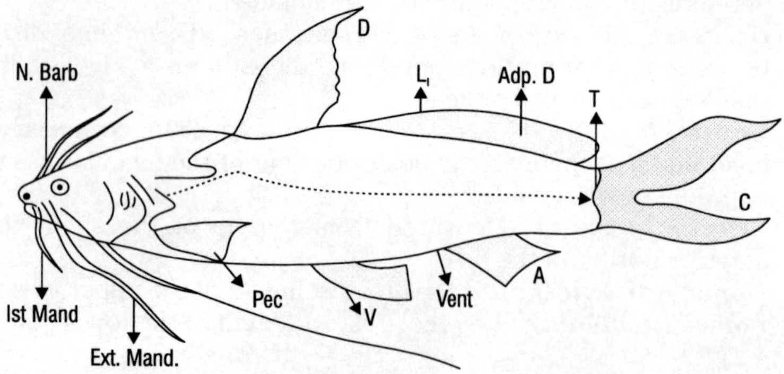

Fig. 1.1: A Catfish showing external morphology

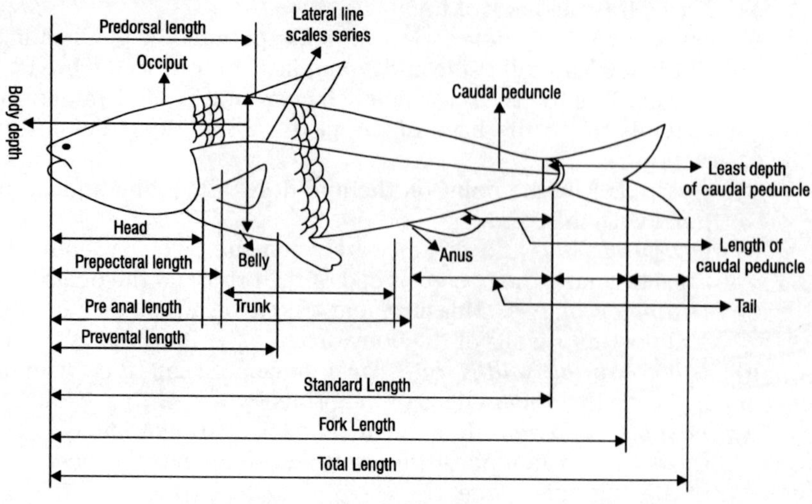

Fig. 1.2: The fish showing various body measurement.

Meristic Character

The characters depicted by numbers are considered as meristic characters. These includes the number of branchiostegal rays, fin rays, scales, barbells, gill rackers etc. Standard abbreviations are used as follows.

for example

A = Anal fin.
B = Branchiostegal rays.
C = Caudal fin.
D = Dorsal fin.
Ll = Lateral line of perforated scales.

Lr = Lateral row of un perforated scales.
Ltr = Lateral transverse rows of scales.
O = Adipose fin.
P = Pectoral fin.
V = Ventral (Pelvic fin).

Branchiostegal Rays

These are curved bony rods attached to the hyoid comua and add to the expense of the opercular gill cover. A definite number of *branchiostegal rays* are present on either side of ventral aspect of the head, and are designed in the fin formula as B. III or B. XI.

Fin Rays

The fins normally have undivided and divided rays. An oblique (/) stroke is used to separate the two kinds of rays in the fin formula. The undivided rays are designated by Roman numerals and the divided rays by Arabic numerals. As for instance, dorsal fin has 3 undivided and 7-8 divided rays it will be depicted as D. III/7-8. If there are two dorsal fins, the first and the second dorsal fins are shown as D_1 VI-VIII. D_2 1/10-15, which means that the first is composed of all spines and the second of one undivided and 10-15 divided rays. Sometimes, there is an adipose dorsal fin also, such as in siluriformes (Cat fishes) and salmoniformes (Trout), which have no bony rays. In such cases the formula is written as, D-1/7-8/0. Usually in dorsal and anal fins, the last divided ray is longitudinally divided up to the base and therefore, both the branches should be counted as one.

Pectoral	:	Fin (1/8) has fin rays, out of which the first one is spinous and the rest 8 are soft ones.
Ventral	:	Or pelvic fin (6) possess 6 fin rays of one type only.
Anal fin	:	11 (2/9) consists of 11 fin rays out of which 2 are spinous and the rest 9 are soft ones.
Caudal fins	:	(17) All 17 fin rays are of single type only.

Scales

A row of perforated scales between the upper corner of the gill opening and the base of caudal fin, in the longitudinal series is counted and their number is shown after the abbreviation L.1. If the lateral line is absent or interrupted, the number of scales in the longitudinal series is denoted after L.r. The scales in the transverse series are counted between the base of the dorsal fin to the row of scales in the mid ventral line of the body and their number is written after the abbreviation L.tr. for example L. tr.21, which may also be written as 8½/12½ or 8/1/12. In the former '1/2' scale is show on both sides of the lateral line while in the later '1' scale shown is that of the lateral line. Here an oblique (/) stroke separates the rows of scales above and below the lateral line.

Other scale counts though not represented in the fin formula, are usually taken into consideration while describing or identifying a species. There are:

1. Pre-dorsal scales counted anterior to the base of the dorsal fin.
2. Number of rows of scales counted between the lateral line and base of the pelvic fins.
3. Number of rows of scales counted around the caudal peduncle.
4. In special cases, such as in *Schizothorax*, the number of scales along the anal sheath are counted, which are more or less stable with a range of variation within species and thus form a good specific character to separate species of the same genus and closely related taxa. Thus the fin formula can be formed as "D 1/7/0; P1/8; V.6; A 11 (2/9) C.17; barbels 4 pair".

But in the above example, the abbreviations like L.1, L.r. and L.tr. are not used as Siluroid fishes do not possess scales. Scaly fishes have a slightly different fin formula *e.g.* "D.13/12-13; P.17; V.1/5; A3/8: C15; L.l 46; L.tr.6/17"

This formula indicates that dorsal fin is single but possess two portions with different type of fin rays. The first part of the fin consists of 13 fin rays, of one type say, spinous and second part possess 12 to 13 fin rays of other type; say branched ones.

Barbels

These are thin, very small or very long, thread like sensitive extensions of skin present around the mouth in carps and cat-fishes and their number, presence or absence usually provide a taxonomic basis to identify a particular group of a fish, although they are subjected to great variation according to the ecological conditions. These are named after the part from which they spring as nasal, from the region of the nostrils; rostral, from the snout; in maxillary, from the upper jaw and mandibular from the lower jaw. Their number varies from 1 to 4 pairs.

Gill Rakers

The number of spine like projections on anterior side of the first gill is counted, as they are specific to certain species.

Example of some meristic counts are presented below

(i) Fin formula: The fin ray formula expresses the count of fin rays in different fins of the body of a species of fish. Spines counts are expressed by roman numerals while the soft ray counts are expressed by Arabic numerals. Spines are the rays that are single shafted. Soft rays are the rays that are bilaterally paired, segmented and flexible. In the case where both spines and soft rays coexist, the roman and Arabic numerals are used to give their counts in the same order. Often capital and small roman numerals are used to distinguish between strong and weak spines. Some cases are cited below to explain this.

a) D II 18 means that the dorsal fin has two strong spines and eighteen soft rays.
D ii 18 means that the dorsal fin has two weak spines and eighteen soft rays.
D 12-14 means that there are only soft rays present in the dorsal fin whose number may vary from twelve to fourteen.
D iv-v 14-16 means that the number of weak spines in the dorsal fin varies from four to five and that of the soft rays from fourteen to sixteen.
D_1 I 4 means the first dorsal fin has one strong spine and four soft rays.
$D_2$5 means the dorsal fin has five soft rays alone.
D ii 15/0 means the first dorsal fin has two weak spines and fifteen soft rays but the second dorsal fin is an adipose fin (without rays).
Dx+I 26-27 means that the dorsal fin has a spinous and a soft part, the former having ten weak spines but the latter having one strong spine and twenty six or twenty seven soft rays.

b) A+C100 means that the anal fin is confluent with the caudal fin and the total count of the combined fin is 100 (soft rays only). For example:

c) Fin formula for *Clarias batrachus*
According to Day 1878 D 62-76, P1/8-II, V_6, A 45-58, C 15-17.
According to Thakur 1978 D 59-66, P1/8-9, V 4-7, A 42-52. C14-16.
According to Talwar and Jhingran (1991) D 70-76, A 45-58. PI 8-11, VI 5.

(ii) Scale Counts: Scale counts with particular reference to the lateral line are commonly used in taxonomy. Some examples are given below.

i) LI 30 means that the numbers of scales on the lateral line of trunk is 30.

ii) Ltr. 16 indicates that the number of the lateral line transverse scales *i.e.* falling on a line from the origin of the dorsal fin to the origin of the ventral fin is 16.

iii) Ltr. 6/1/4 would mean that there are six scales above the lateral line scale and four scales below it.

Descriptive Characters

1. ***Shape of the body:*** The curvature of the dorsal and ventral surface of the body also marks the recognition of the fish.

2. ***Parts of head regions:*** The type of lips, jaws, nostrils, eye lids, presence or absence of pores or tubercles on the snout, rostrum and other parts of the head, position, nature of anal pore, genital papilla and extent of adipose dorsal fin, if present, includes some of the descriptive features.
3. ***The nature of gill membranes:*** They may be free or united with the isthmus. The isthmus may be narrow or wide and its ratio with other dimensions of the head form good taxonomic character.
4. ***The orientation of paired and unpaired fins:*** The position of pectoral and pelvic fins (Paired fins) differ in different groups of fishes. The pelvic fins may be abdominal (in the region of abdomen), thoracic (in the region of thorax or below pectoral fins) or jugular (in front of one pectoral fin).

   The origin or commencement of the dorsal, pectoral, anal and pelvic fins and their relation with one another is taken; as a taxonomic character for the identification of fishes. The shape of the fins on the other hand is definite in each group of fish and hence important as an identifying character.
5. ***Sexually dimorphic characters:*** The special morphological modifications of different nature in males and females should be observed, because they are very important in identification of species. Moreover, sex differentiation is of the most importance in fish breeding, particularly in artificial or induced breeding for cultural purposes.
6. ***Colouration of body:*** The colouration of body and fins is also sometimes diagnostic in many groups of fishes, although it is likely to vary with different ecological conditions.

Key to Fish Identification

Different kinds of keys for the identifications of fishes are employed, such as the running key, dichotomous key, open or closed keys etc. But it has been seen that the dichotomous key is more helpful for both, a beginner as well as an advanced student of fish. Each key mentions such characters as differentiation which closely separates the different taxa from one another. For a preliminary sorting of a sample of fish into different groups (order, families, genera and species etc) the key to their identification is normally employed. For further details, the description of those taxa is then consulted.

For easy identification of fishes along with details, step-wise character, clear and good illustration should also be consulted from standard books and the specimen in question should also be compared with the labeled, identified preserved specimens of the collection.

Chapter 2

Origin and Evolution of Fishes

INTRODUCTION

The history of life, like human history, has been marked by certain great development rising above the general level of events that collectively make up the record.

Before dealing with general problem, it is of primary importance to make it clear, what is meant by a fish. The word fish was used to include any animal living in water. A definition appears in all older dictionaries.

Although convenient, this can hardly be described as scientifically accurate, including, as it does, such diverse organisms as whales, seals, salmon, oysters, cuttle fishes, star fishes, jelly fishes, sponges and other creatures that differ from each other even more widely than do true fishes from mammals. The aquatic animals just mentioned however, all fall naturally into two main categories in respect of one important feature, those with a vertebral column or back bone and those with none. Man has a back bone and so have all the mammals, birds reptiles, amphibians and fishes but all the other have no back bone.

Therefore, a fish may be defined as "A vertebrate adapted to purely aquatic life propelling and balancing itself by fins and obtaining oxygen from water for breathing purpose by means of gills".

ORIGIN

The phylogenetic relationships and the probable evolutionary stages resulting in the most perfectly adapted orders.

The first vertebrates known in the fossil record are indicated, by scales that have been found in Ordovician fresh sediments in Colorado. These remains are fragmentary and give no clue to the nature of animals represented by them, yet they are sufficiently well preserved to be studies.

Jamoytius and Thelodus from the middle Silurian of England may be considered as a representative of the earliest chordates and had v-shaped myotomes and continuous median and lateral fin folds.

The craniates may be considered to have evolved from Jamoytius like ancestors, some fragmentary fossils of the earliest known fishes are those of the ostracoderms from the Ordovician age, suggesting that these jawless fish-like vertebrates arose about 400 million years ago. However, by the Silurian time the Agnatha and the Gnathostomata were very well established. Jamoytius and Theodus are found together in marine sediments. If we use the word "fishes" in broad sense, we should say that the first vertebrates were fishes but actually the first fishes were primitive jawless animals known as Agnatha. A word compounded from Greek roots or words (A=without *gnatha*=jaws) from these early agnatha us animals the Gnathostomata jawed vertebrates originated first the fishes and later on other vertebrates. Table 2.1.

EVOLUTION OF FISHES

One of the great events or revolutions in the history of the vertebrates was the appearance of the jaws. The importance of this evolutionary development can hardly be overestimated, for it opened to the vertebrates new lines of adaptation and new possibilities of evolutionary advancement that expanded immeasurable potentialities of these animals. The jawless vertebrates were definitely restricted as to their adaptations for different modes of life and it is possible that the ostracoderms of late Silurian and, *Devonian* times had explored and virtually exhausted the evolutionary possibilities and the ecological niches open to animals of this type. These jawless vertebrates are classified into two classes.

1. Ostracoderm (Extinct)
2. Cyclostomata

Now taking a general survey chiefly discussing those facts which have some significance from evolutionary point of view that to what extent these jawless vertebrates are responsible for origin of modern fishes.

Ostracoderm (Agnatha)

The jawless armoured ostracoderms flourished during the late Silurian and Devonian periods. They are considered to be related to the cyclostomes due to resemblance in the structure of the brain, ear and blood vessels. The ostracoderms varied in length from a few inches to about a foot and most of them were protected by dermal bony plates but a few had a reduced armour, the bony covering might have served as defensive mechanism against scorpion like eurypterids of the Silurian age. Table 2.2.

Table 2.2:

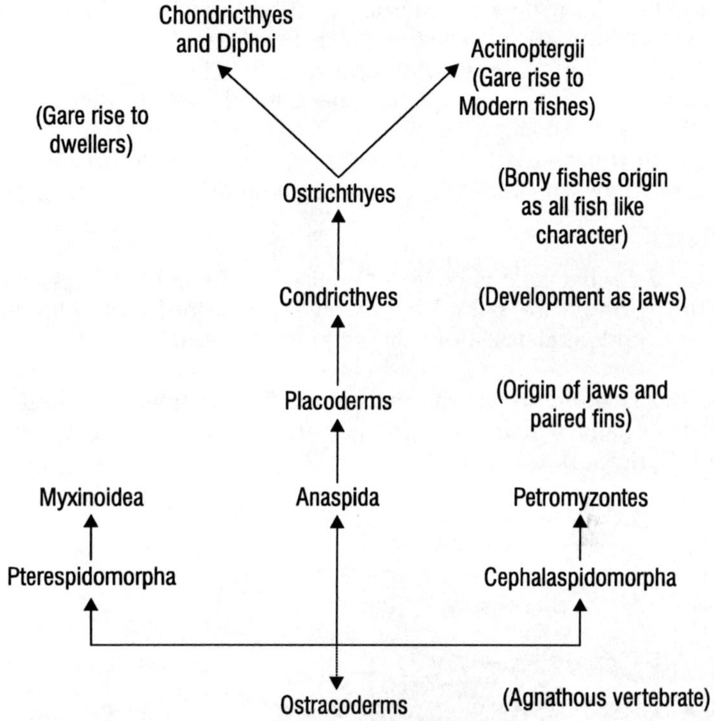

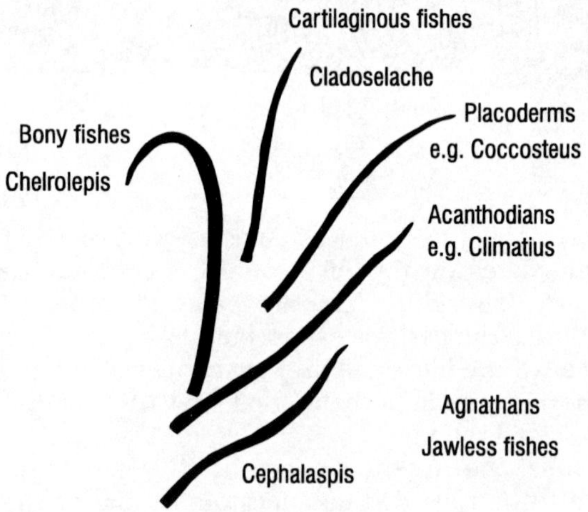

Fig. 2.1: Evolution of primitive fishes as prepared by Lois M. darling

In fossil records, we find that oldest and most primitive of fossil vertebrates (Jawless) were found in Ordovician and Silurian deposits, among the best known ostracoderms were cephalaspis, which showed a striking diagrammatic way to the general pattern of fish.

This was strikingly fish like in appearance. The trunk and tail, are very much, like those of a fish. The caudal and dorsal fins are comparable with those of a shark. The forward part of trunk is covered with transverse bands like scales, which grade off into more typical scales towards tail region are seen in modern fishes. (Fig 2.1)

Cephalaspis

The head of cephalaspis is covered with a head shield. Attached to it each side of body are paired appendages, that look something like paired fins, endoskeleton bony or completely ossified as in modern fishes.

Another striking peculiarity in this primitive fish was the presence of electric organs which are also seen in modern electric fishes ex. Torpedo, Gymnotus etc. (Fig. 2.2)

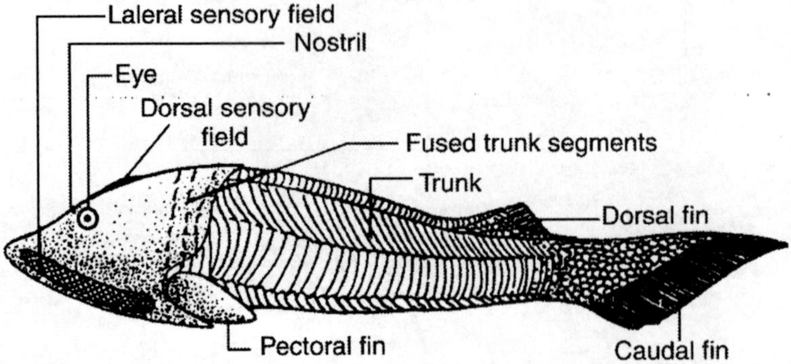

Fig 2.2: *Cephalaspis*

Another ostracoderms, such as Anaspis were small headed fish like forms with plates on top of the head, somewhat suggestive of those of modern fish skull. They are believed to have been swift swimming form like the earliest fishes, but lacks jaws. They could not have captured anything but small prey that could be taken in by round jawless mouth. It seems improbable that first true fishes are derived from general anaspids stem.

On the whole, the question arises as to whether ostracoderms themselves are to be regarded as ancestors of any of the swimming vertebrates yes, there is no doubt actually in regarding this fact because, a great deal of recent study of group to believe in the words of Newman "some of them" anaspids pteraspidomorphi and cephalaspidomorphi are not far from the lines of descent of respectively the hag fishes and lampreys. For a more detailed study of the characteristics of limbless jawless level of vertebrate evolution we shall now proceed to an examination of existing cyclostomata.

Cyclostomes

The cyclostomes are surviving limbless jawless vertebrates sometimes called round mouth eels. The class consists of two subclasses that are not at all closely related, but one doubtless end products of two lines of evolution going back to two different orders of ostracoderms, being as a rule of evolution, they must be highly evolved than ostracoderms.

But it is to be noted that cyclostomes are the only group in animal kingdom which are popular and outstanding in the presence of primitive characters, they have acquired these primitive characters only to adapt themselves to their special parasitic mode of life. In the words of Newman we may again say "sub-class myxinoidea is regarded as a degenerate offshoot of pteraspidomorphi and sub-class petromyzontidae as a slightly less degenerate descendent of cephalaspidomorphi.

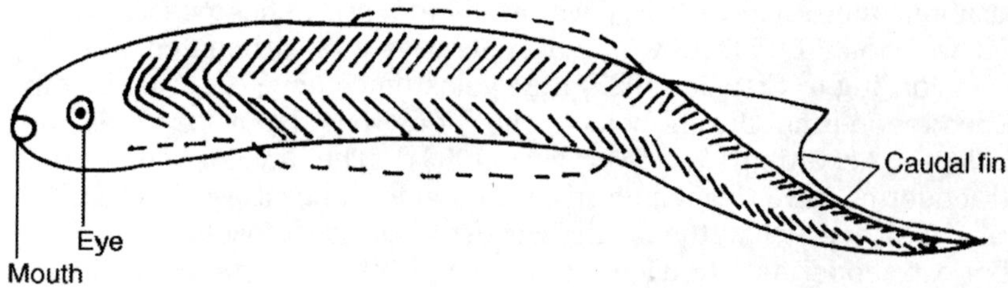

Fig 2.3: *Jamoytius*

However, there remains no doubt that cyclostomes are most primitive of swimming vertebrates. In a popular and general sense the cyclostomes are usually regarded as "fish" in their methods of respiration by gills. This resemblances reveals the fact that cyclostomes have also to do something so as to give rise to the origin of true fishes.

Now, no account can serve better than taking in comparison between certain characters of cyclostomes and fishes that help in recognizing the general problem.

1. As cyclostomes are totally parasitic, there is no need of paired limbs, only a single dorsal fin, confluent with caudal fin is seen. The caudal fin in continued ventrally to form ventral fin. These

median fins are also seen in modern true fishes. Alongwith these median fins, the true fishes show slight advancement in having paired fins also.

2. Again being as parasitic, the cyclostomes do not need any device for protection as a result of which they lose their exoskeleton. But in modern fishes body is covered by scales for protections.
3. Cyclostomes are very primitive in having no jaws and this primitive character is again acquired due to their parasitic mode of life. In petromyzon mouth is round and suctorial and is provided with teeth for grasping the flesh. The suctorial nature is meant for attachment. But in true fishes there are well developed jaws.
4. Lamprey differs from all other true fishes in the fact that gills do not open directly into enteric canal but into respiratory tube opening in front into the buccal cavity. The respiratory organs are typical gill pouches leading to branchial apertures. While in myxine tubes leading from gill pouches unite together so that there is only one single branchial aperture.

In true fishes there are five gill pouches which open directly into enteric canal and are covered by bony flap operculum. During the course of discussion which has been mentioned above, we reached to the conclusion about the level of jawless fishes without getting any hint about the origin of true jaws but as modern fishes possess, true jaws, we should find from where the true jaws were originated.

According to Stensio (1927) the cyclostomes have described from the ostracoderms during which the bone has been replaced by cartilage. According to Svertsor (1939). The cyclostomes and ostracoderms have developed from a common ancestor. There is no-doubt, however, that the living cyclostomes are closely related to fossil ostracoderms. The view of Stensio (1927) can be schematically represented as below.

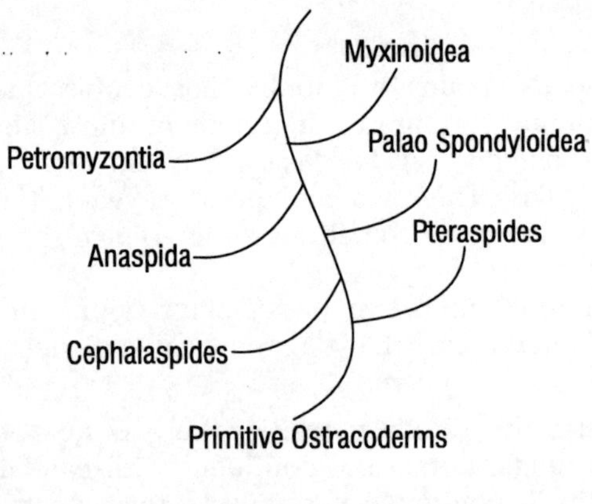

Placoderms

The early jawed fishes arose from the Agnatha in which the small terminal mouth was extended posteriorly and the first visceral arch was covered into a primitive jaw. The placoderms are the earliest gnathostomes and appeared during the upper Silurian.

The acanthodians are the earliest placoderms and superficially resemble the sharks. They possess characteristics spines to support all the fins except the caudal, which is heterocercal. The body was covered by small, flat bony scales and had not developed cephalic and thoracic bony shields found in all other placoderms. These fishes show a mixture of primitive and specialized features and closely resemble the ancestral stock that gave rise to the bony fishes. Fishes belonging to the orders coccosteus and pterichthyodes were covered with bony shields on the head and the anterior part of the body. The stegosaurus were less heavily armoured and superficially look like modern rays in the shape of their body and pectoral fins. However, none of these groups is in the main evolutionary line of the higher fishes. *Palaeo spondylus* is also considered to be related to the placoderms.

Evolution of the Cartilaginous Fishes

The cartilaginous fishes, Chondrichthyes are very likely the descendants of some placoderm ancestor possess jaws and paired fins and include shark, rays, chimaeras etc. They appeared in the mid Devonian period and soon became well established continuing up to the present time. They are primarily marine fishes and possess a cartilaginous skeleton which is probably a secondary feature and does not represent a primitive condition. Sharks are generally considered to be primitive than bony fishes, but they appeared at a later stages than the bony fishes. Cladoselache is one of the earliest sharks that lived during upper Devonian. It had a typical shark-like body heterocercal tail, two dorsal fins, pectoral and pelvic fins. The pectoral fins were very large and were simple extensions of the body wall. The skull was amphistylic. From such an ancestral type, the cartilaginous fishes evolved along with several fishes. The most primitive of these were the pleuracanthus sharks that appeared during the carboniferous and are exemplified by the genus pleuracanthus. Like the cladoselachian they also had amphistylic jaw suspension but were specialized in some respects. The body was shark like with a diphycercal tail and long dorsal fin. Paired fins were of the archopterygeal type and a long spine projected backward from the skull. During the Mesozoic era the advanced sharks appeared in which the skull become hyostylic providing more freedom of the jaws which is a characteristic feature of the advanced sharks. From the Mesozoic times of sharks evolved along two lines, giving rise to the modern shark and the rays. The sharks are active fast swimmers and predaceous fishes. They possess a torpedo like body, strongly heterocercal tail and paired fins with narrow bases. The skates and the rays constituting the other evolutionary line become specialized for bottom living habit. In these fishes the pectoral fins are greatly enlarged and fused with the body. The tail is narrow, whip-like and the gill sils are ventrally placed. Some genera, Heterodontus, Hexarchies and Chlamydoselachus occupy an intermediate position between the primitive cladoselachian and the advanced sharks and may be called "living fossils". (Fig. 2-4) and (Table 2.3).

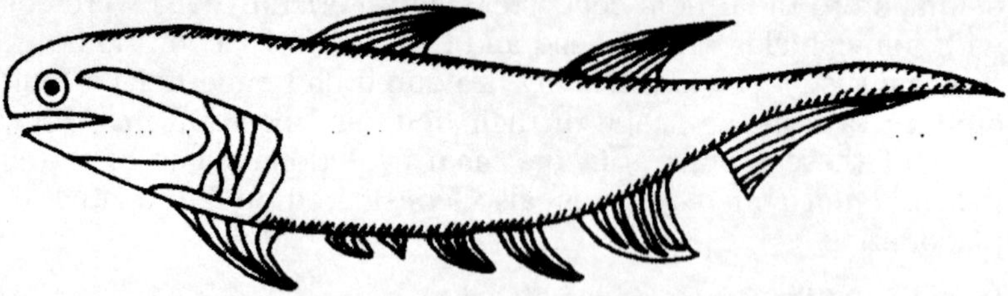

Fig 2.4: *Climatius*

Evolution of Bony Fishes

The bony fishes (Teleostomi) appeared in the middle Devonian and were probably derived from some shark like fishes. The early shark was evolved in the fresh water in large rivers so the bony fishes also arose in fresh water. Sharks has always been exclusively gill breather and never developed any equipment for breathing air. All the bony fishes however, develop lungs and became dominant vertebrates of the fresh water. However, as four major group of these fishes were present during the Devonian. Some authors believe that the bony fishes possibly had their origin in the Ordovician more than 400 million years ago.

1. **Diponi**: Diponi arose by the middle of the Devonian and certainly derived from Primitive Crossopterygii. They are now regarded as a somewhat specialized group that early became well adapted to life in temporary water, but still remain purely fishes. A majority of them became extinct long ago, a few species have survived in isolated region upto the present. The earliest lung fishes of which dipterous is a typical example. The body is fusiform, internal nostrils and a heterocercal tail.

Table 2.3: Genealogical tree of cartilaginous fishes

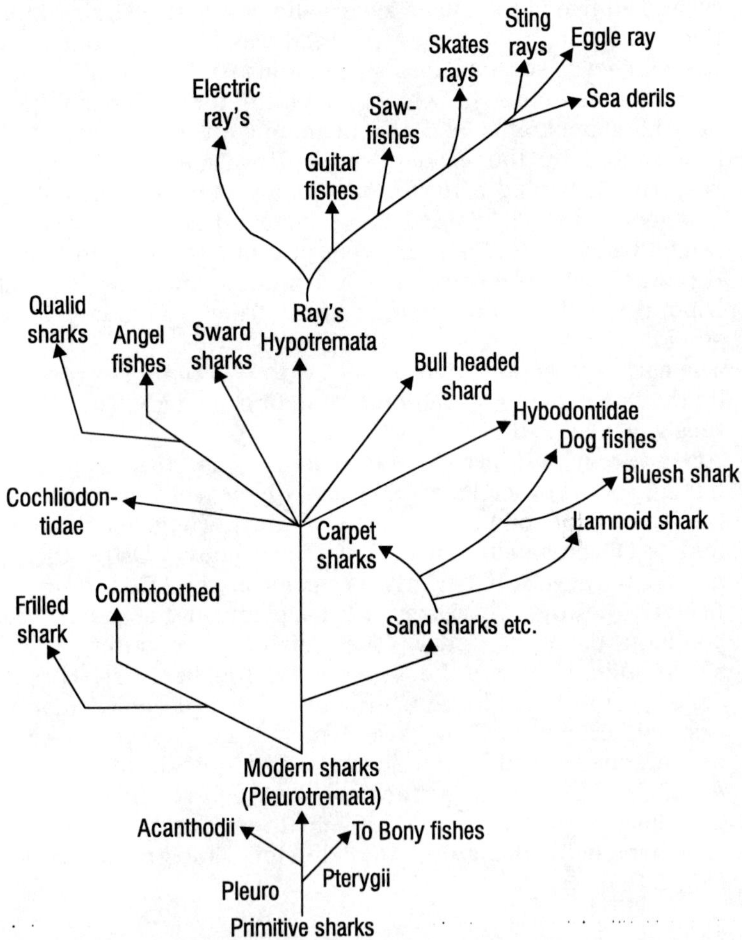

It possess certain specialization in fins, scales and the bones of the skull gradually gave rise to the later Diponi, which are now represented by Protopterus, Lepidosiren and Neoceratodus which become adapted for life in rivers that are likely to dry up.

The modern view as to the reason why the lung fishes show some trends toward land vertebrates is that they and the Amphibia are both ascendants of a common ancestral stock, the primitive Crossopterygii, which themselves had already started the development of adaptations favourable for land life.

2. **Crossopterygii. (Lobed finned fishes):** These fishes also appeared in the middle Devonian, but they declined during the

carboniferous period and only a few of them persisted in to the Mesozoic era. They are the fishes that pushed farthest up the rivers and live in the inland swaps and ponds that drained into the river. In swamps and ponds, the food was abundant but the water was subject to seasonal changes, tending to dry up during the long rain less summers in which lobed fin fishes having good air breathing apparatus and the beginning of land legs. They are represented by the genus Osteolepis which closely resembled Dipterus in having a fusiform, strongly heterocercal tail, lobate archaeopteryx gel fins and heavy cosmoid scales. The skull was ossified and internal nostrils were present. There is no doubt that Dipterus and Osteolepis had a common ancestor so that the Diponi and the Crossopterygii arose from a common ancestral stock.

The early crossopterygians gave rise to two branches represented by the sub orders *Osteolepis doti* and coelacanth. The Osteolepis thus gave rise to the tetrapods.

3. **Actinopterygii (Ray finned fishes):** That first Actinopterygii belong to the order Palaeoniscidae which were characterized by the possession of heavy rhombic scales. They appeared in the middle Devonian along with the other primitive Osteichthyes, but at first they were not nearly as numerous as lobed fin and lung fishes. From these early ray fins have sprung the majority of the common fishes of today. The Teleostei, constitute the most successful survivor of the ancestral ray-finned fishes. These early fishes were of small size with heterocercal tail and incompletely ossified vertebrate. They have ctenoid scales and their pelvic fins are migrated towards anterior side. The notochord was strongly developed. The later, actinopterygians evolved from Chriolepis like ancestors and passed through three stages of development represented by the group Chondrostei, Holostei and Teleostei. (Fig. 2.5)

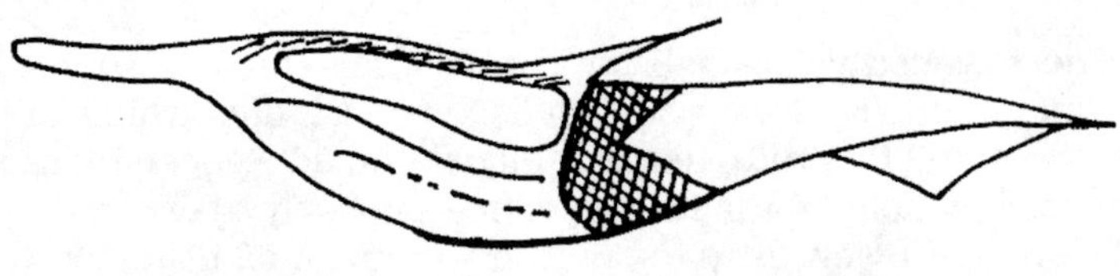

Fig 2.5: *Pteyasipa rostuata*

4. **The Brachioptertgii:** They were represented by two surviving genera Polypterus and Collomoichtyes arose from the palaeoniscoids and are confined to tropical African rivers. They

are "living fossils" and possess both primitive and specialized features. They are usually classified as an order of Actinopterygii but show some important resemblances with the Crossopterygii while exhibiting differences from the typical Actinopterygii and they also differ from Actinopterygii in having an air bladder attached ventrally to get rather than dorsally. Hence, some authors are now in favour of treating them as a separate sub-class. (Fig. 2.6)

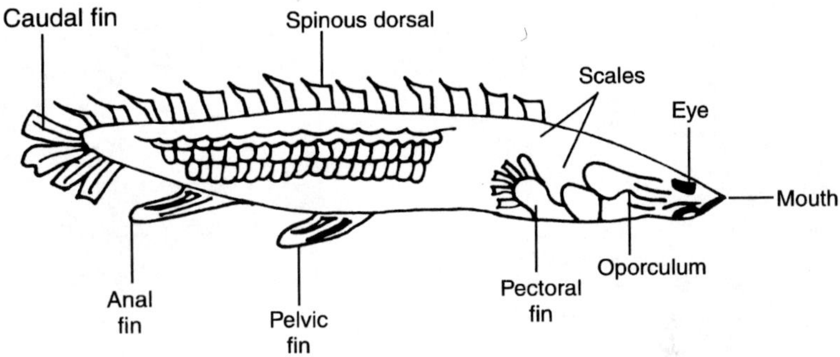

Fig 2.6: *Polypteuels victus*

The palaeoniscoids also gave rise to the Sturgeons represented by Acipenser. Which possess the spiracles but the heavy scales have been lost. Both Polypterus and Acipenser and the chondrostean fishes during the Triassic. The palaeoniscoid ancestors gave rise to certain progressive chondrostean fishes represented by the genus red Fieldia. They show a reduction of the upper lobs of the heterocercal tail, less heavy scales and shorter jaws. They can be considered intermediate between the Chondrostean and Holostean fishes.

At the end of the Triassic, the primitive chondrostean were gradually replaced by the Holostei, which were more specialized than ancestors. The Holostean are now represented by two genera Arnia and Lepisosteus both of which possess a vascular lungs for respiration.

The Holosteans were gradually replaced by the teleosts and was represented by the genus Leptolepis which appeared in the Jurassic and intermediate between Holostean and the teleost fishes. In teleost the scales lost the layer of genoine and became thin and transparent. The caudal fin is completely homocercal. The vascular lung is modified into swim bladder. The internal skeleton is ossified with specialization in the skull. The teleosts were very well established and have radiated into a number of adaptive lines. A number of teleosts live in the fresh water to which they have returned from the sea and represents the most successful group of the fishes. (Table 2.4).

Table 2.4: Genealogical tree of Bony Fishes

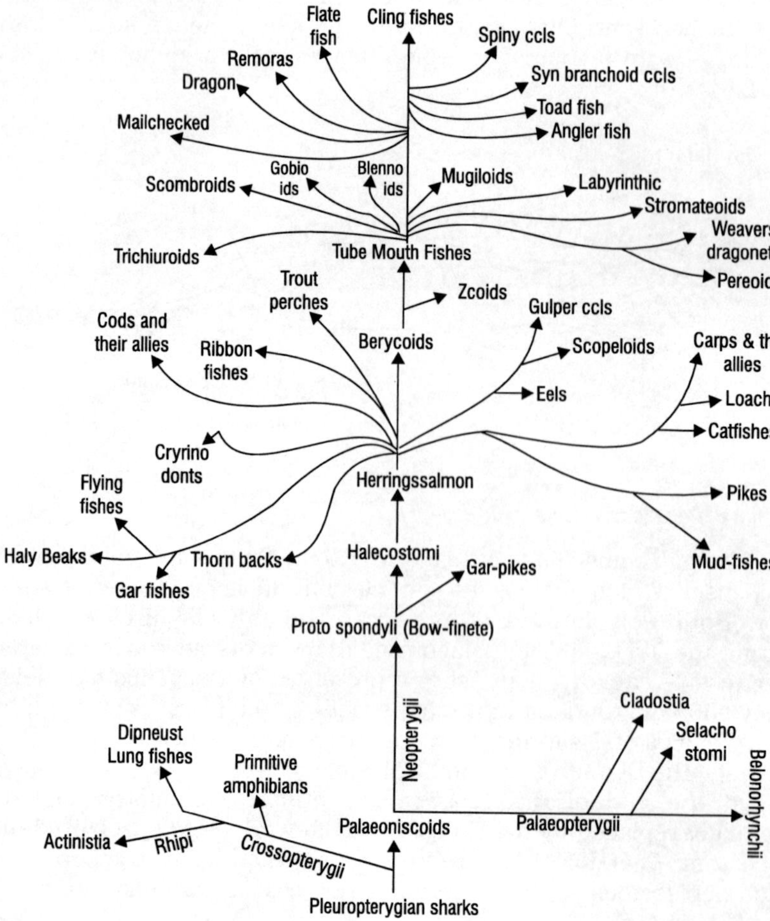

CONCLUSION

The placoderms lived in the Devonian period, were characterized by the presence of jaws, well developed fins and reduced exo skeletal armour. The Ostracoderm disappeared by the end of the Devonian period. The oldest fish fossils have been unearthed, from the rocks of the Ordovician periods some 5, 000 million years ago. The chondrostean were first fishes to evolve and they were dominant during late Paleozoic period and were replaced by Holostean in the

Mesozoic era which were ultimately replaced by the teleosts.

The first Devonian crossopterygian represented by Osteolepis from the basal stock from which crossopterygian evolved. The evolution took place in two general line, one being represented by the sub-order Rhipidistia and other by the order Coeloncanthini. The placoderm are supposed to be related of the early sharks.

Chapter 3

Algal Blooms

INTRODUCTION

Sudden appearance of blue green algae as (blooms) or breaks in association with fertilization. These blooms along with different types of aquatic weeds liberate good amount of carbon-dioxide absorbing the available oxygen leading to its depletion during night. It is reported that the carbon dioxide at a concentration of 20-30 ppm may be responsible for large scale fish mortality. Hence, suitable measures have to be under taken to eradicate these unwanted *aquatic flora*.

The basic pattern for the metabolism of a fresh water algae is purely an autotrophic one involving the fixation of carbon dioxide by photosynthesis and the elaboration of cell constituents from chemically simple starting materials. Assuming a maximum efficiency in the metabolic process carbon dioxide derived from respiration would be the whole waste product of the carbon assimilation. However, these two features of the basic patterns do not always apply. Many algae can utilize organic substances such as sugars or organic acids to excrete organic compounds in water in which they grow.

DEFINITION

Certain planktons are very quick to grow in favourable conditions, which form the *blooms* or tides. These blooms occur in fresh, brackish and salted water.

GENERAL ACCOUNT OF ALGAL BLOOMS

Algal blooms are invariably caused by unicellular and filamentous algae and may impart any colour to the water, like green, yellowish, brown, reddish brown bluish green etc.

The group of algae which are responsible for the development of blooms include Chlorophyceae (green algae) Xanthophyceae, Bacillariophyceae, (diatoms), Chrysophyceae (golden-brown, flagellates), Cryptophyceae, (biflagellate), Dinophyceae, (dinoflagellates) Euglenophyceae and myxophycea (blue-green algae). During rainy season the blue-green, some green and some golden brown algae and the diatoms became so abundant that the water in the ponds, lakes and reservoirs becomes cloudy and assume a yellowish, greenish tinge. Sometimes a floating, yellowish, green scum may develop on the surface of water. Popularly we call these manifestations of algal growth as water blooms.

GROUPS OF ALGAL BLOOMS

1. **Phytoplankton:** The organisms are exclusively of plant origin and are thus autotrophs, belonging to the first trophic level (Producers).
2. **Zooplankton:** The organisms are exclusively animals and are therefore heterotrophs belonging to the second trophic level. (Primary consumer).

Both are in one way very dangerous to fish farms because in each case the respiration at night will cause serious oxygen depletion in water which may in turn cause heavy fish kills.

TYPE OF ALGAL BLOOMS

Temporary Algal Blooms

The common temporary bloom forming algae in the pond water include euglena and Chlamydomonas, (Euglenophyceae) Volvox, (Chlorophyceae) and Peridinium (Dinophyceae) and their respective colorations are reddish brown, green and brown. The appearance of these algae in high densities may be avoided by the application of fertilizers and supplementary feeds for some times.

Persistent Algal blooms

When blue green algae such as Microcystis, *Anabaena* Oscillatoria, Nostoc, and Spirulina appear in profusion, it may cause a surface scum, which in turn may lead to fish mortality, Microcystis and Anabaena, possess vacuoles, filled with oil or gas which expands warm water, lifting the algae to the surface. The penetration of light to deeper waters is greatly affected. Excess levels of nitrogen, phosphorus, carbon dioxide and organic matter are considered to be responsible for the fast propagation of these algal species. The increase in surface temperature and pH > (09) influence the formation of persistent algal blooms.

BLOOM FORMATION

Bloom formation only occurs when there are numerous algae present in a lake, and the resulted visual spectacle represents the sudden concentration of high to more dispersed population rather than the formation of a new one.

A study of bloom formation of Anabaena circinalis indicates that photosynthetic activity and internal vacuole formation are intimately inter-related in such a way as to control the buoyancy of this algae and hence its potentiality to rise and aggregate as a bloom.

The internal vacuoles have constructed walls and are gas filled, the volume of the cell they occupy varies according to the rate of growth of the algae, thus in the warm period leading upto the time of bloom formation in May, growth of the alga was rapid and total vacuole volume was 1:1 percent of the cell volume with some evidence that the production and erection of gas vacuoles lagged behind. The rate of cell division, once growth had ceased due to the limitations of nutrients, the gas vacuole volume rose to 3.8% of the cell volume and the cell become more buoyant. At this stage Anabaena was in the epilimnion down to a depth of 6 meter, and was generally very buoyant but was also undergoing regulatory movements in a vertical direction. Thus, when filaments of the alga sank out of the illuminated zone, the cells suffered a reduction in their rate of photosynthesis. This led to a decrease in turgor pressure in the cells because there was less manufactured sugar.

In turn, an increase in the relative vacuole volume an increased buoyancy resulted, bringing them back in to the region of maximum photosynthesis. The vertical movements in both directions were enhanced by turbulence but once this was reduced by the sudden onset of calm fine weather. The generally buoyant nature of the population brought it to the surface, where it become lodged in a film under strong illumination which was unsuitable for satisfactory photosynthesis. Under these conditions turgor pressure of the cells decreased and relative vacuole volume increased, thus strengthening the buoyancy of the whole Anabaena film light breezes, later deposited the film at the edge of the mere. (Fig. 3.1).

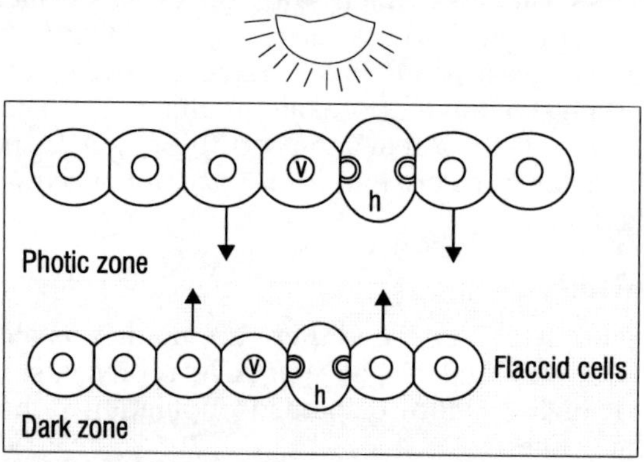

Fig 3.1: Regulatory movements of *Anabaena circinalis* in relation to differing cell turgidities and buoyancies.

ALGAL BLOOMS TOXINS

Some algal species have the capacity to produce and eventually release toxic metabolites. These can accumulate in the environment to such an extent that they cause the death of fish and other aquatic animals.

Exotoxins

Exotoxins are released by certain algae during their life, and the flagellate Prymnesium parvum, a member of the Chrysomonodinae, exemplifies this type-

Prymnesium

Prymnesium is essentially a marine alga, but it flourishes in brackish water in Europe and Israel. The lethal effect of the Prymnesium exotoxin is exerted on all gill breathing animals including both molluscs and fishes. The main symptoms which follows the immersion of fish in the gill toxin is a loss of selective permeability in the gills; toxicants, which are not lethal to normal. Fish under these circumstances cannot resist concentrations of copper and other ions in the environment which are normally sublethal. In addition to this effect, (called ichthyotoxic) the toxin has a direct effect on cells, causing disintegration of both blood corpuscles of the blood system (haemolytic effect) and of nucleated cells in general. The chromatographic analysis of the Prymnesium toxin indicates, that it consists of six components, each showing a different ratio for its relative ichthyotoxic and haemolytic activity. The full chemical nature of the components is not known at present, but there are indication that proteolipids are present containing proteins, polar lipids and phosphate. Light is essential for the production of toxin in the Prymnesium cells but the toxin itself is inactivated by the exposure to light or ultraviolet radiation. The Prymnesium is potentially most dangerous in conditions of low salinity of the environment. Close to the lower limit at which the cells essentially halophytic one, can survive.

Endotoxin

Endotoxin are also produced by Gymnodinium and blue green algae such as Microcystis aeruginosa. Endotoxin are only released after the disintegration of parent cell. The Microcystis endotoxin, which is possibly a cyclic polypeptide is released when the algal cells are disrupted by freezing and thawing. The Microcystis algal blooms never give rise to mass mortalities of fish or water birds. The endotoxin released from a blue green alga could occur in natural waters or in reservoirs. There have been accounts of particular parasitic bacteria or viruses attacking algal blooms causing epidemics and high mortalities in these and disintegrating algal cells. Under these circumstances the toxin might find its way into the water more readily.

APPEARANCE OF ALGAL BLOOMS

Algal blooms usually make their appearance after treatment of the pond with heavy doses of inorganic fertilizers like urea and superphosphates. The bloom also appears after treatment with weedicides like 2,4-D Sodium arsenite etc.

This discovery, that there are viruses which attack and kill algae is very recent. The viruses is polyvalent in that it can be transferred from one host to another. Anabaena and Microcystis, are apparently not susceptible to viruses and it is problematical at the moment whether viruses do exist which attack them.

DISADVANTAGE OF ALGAL BLOOMS

Algal blooms are responsible for giving a disagreeable taste and smell to the water. They course heavy mortality of the fishes due to choking of the gills, depletion of oxygen and liberation of toxic substances.

CONTROL OF ALGAL BLOOMS

Chemical Method

One ppm copper sulphate solution or copper sulphate at 800 g/ha, packed in cloth bags are immersed in ponds is known to eradicate algal blooms, particularly the blue green algae. It is necessary to apply this chemical once a week and for a number of weeks if needed. Further, 125 ppm Sulphuric acid or 0.5 ppm Sizazine (2-Chloro 4,6 bis-ethylamino-3-triazine) may also control the Microcystis bloom effectively.

The affected ponds may be treated with ammonium sulphate, which precipitates the most important algal growth promoting substance, viz phosphate. Since acidic pH (6-7) disfavors the formation of algal blooms, carbon dioxide may be introduced in the contaminated ponds which certainly brings down the pH of water.

Biological Methods

The algal blooms can also be controlled by introducing the duckweed, lemna, which forms a shade over the algae and prevent their growth. Alternately algae feeding fish such as silver carp may be introduced. Algal blooms can also be controlled by introducing viruses which are known to prevent the photo assimilation of carbon dioxide.

EFFECT OF ALGAL BLOOMS ON FISH

The cells of these blooms choke the gills of fish leading to death. Further, the fish mortalities are said to take place due to the release of toxins during peak photosynthesis, depletion of oxygen liberation of toxic substance in association with the man decay of these algal blooms.

CONCLUSION

Recent investigation at clean lake, California, have suggested nitrogen fixation by blue-green, algae. It contributes 43% percent of the nitrogen available for growth in this water body. The large contribution from algal fixation was suspected for this like before it was investigated in detail, because the nitrogen in the outflow stream greatly exceeds that estimates as being contributed by the inflow sources.

Chapter 4

Aquatic Weeds and their Control

INTRODUCTION

Weed sciences involves the study and control of the more aggressive troublesome elements of world's vegetation and accordingly constitutes parts of such subjects as plant ecology and plant geography. Relatively this new sciences also comprises the other plant science disciplines in addition to certain chemical, zoological & earth science.

In the examination of what precisely meant by the term "weed" we shall consider several aspects. Firstly the origin of the word "weed" is English language, as well as some of the other equivalent languages, together with the development of the weed concept, secondly the contemporary sciences of ecology, cytogenetics and taxonomy offer in the explanation and study of the kinds of plant that are classified as "weed".

Aquatic weeds are the undesirable plants that grow in water and are more harmful than beneficial effective for fish culture. Excess growth of aquatic weeds prevents effective utilization of water and reduce the productivity depending on their habit & habitat.

DEFINATIONS OF WEED

Shorter definition of weed has appeared than that attributed by professor Beal, formerly weed specialist at Michigan State Collage, who defined weed as "A plant out of place". Some more definitions of weed are illustrated below:

1. **A concept of growing in an undesired location:** This is essentially the chief characteristic of the weed definition given in the report of the terminology committee of weed society of America (Shaw 1956) namely a plant growing where it is not desired.
2. **Competitive and aggressive habit:** According to Combell's 1923 the weed is an honest independent competitor for food materials in the struggle for existence. Branchely (1920) defines a weed as "A plant that grows so luxuriously or plentifully that it chokes out all other plants that possess more valuable nutritive properties".
3. **Persistence and resistance to control:** This is inherent in the title of a Gray's Publications of 1879 "the predominance and pertinacity of the weeds".

4. **Consisting often of large populations with abundance: rank and extensive growth:** The native vegetation and weed infestations consist of massive growth, often of a single species or almost of only a few species.
5. **Unwanted weeds:** According to L. H. Bailey & E. Z. Bailey (1941) "An unwanted plant and therefore, to be destroyed and also whose virtues have not yet been discovered".
6. **Spontaneous growth appearing without being sown are cultivated:** Branchey's (1920) defines it-"Any plant other than the crops sown" and Harper (1944) "A plant that grows spontaneously in a habitat that has been greatly modified by human action".
7. **Of high reproductive capacity:** Weeds in contrast to non-weeds generally produce in large numbers. They may also possess highly effective methods of vegetative reproduction.
8. **Unsightly with disfigurement of the landscape:** This is true to indicate by number of definitions such as "A plant which grows wild or is intrusive in cultivated ground" and is not valued for its usefulness or its beauty (Dictionary of American English 1944) by Thomas (1956) "A useless undesirable, often very unsightly plant of wild growth usually found in land which has been cultivated in areas developed by man for specific purpose other than cultivation.

CLASSIFICATION OF WEEDS

Mirashi (1957) in his studies of the hydrophytes of Umred at Nagpur in India has provided a classification into six forms based on their contacts with soil water and air. But actually there are 11 types of weeds.

1. Suspended Hydrophytes

These are root less submerged hydrophytes that are in contact with water only, for example *Ceratophyllum demersum*. (Fig-4.1)

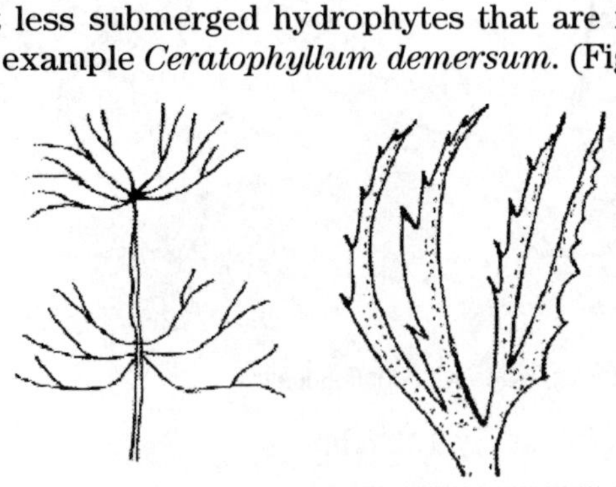

Fig. 4.1: Suspended hydrophytes

2. Anchored Submerged Hydrophytes

These are entirely or for the most part in contact with soil & water. Example *Hydrilla*, *Vallisneria* etc. (Fig. 4.2)

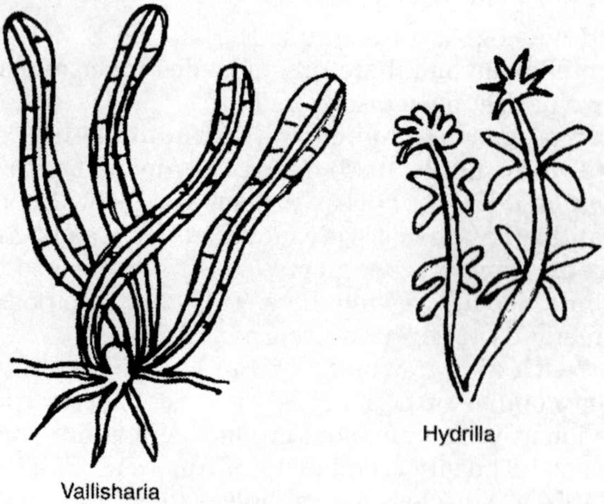

Fig. 4.2: Anchored Submerged Hydrophytes

3. Anchored Hydrophytes with Floating Leaves

They are in contact with the soil, water as well as air, for example *Aponogeton*, *Potamogeton*, *Pectinatus* etc.

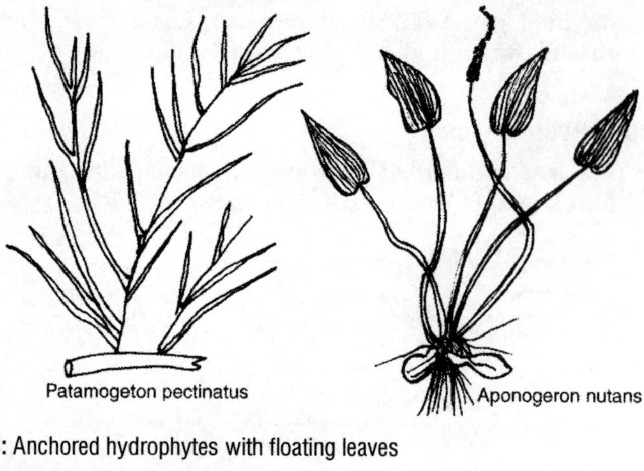

Fig. 4.3: Anchored hydrophytes with floating leaves

4. Emergent Amphibious Hydrophytes

The roots and the lower parts of the stem and in some cases the lower leaves of these hydrophytes are usually submerged underwater example *Hydrophilia*, *Sesbania* & *Limnophyton* etc.

5. Wet Land Hydrophytes

These are rooted in soil & that is usually saturated with water, atleast in the early part of their life Example: *Chara*, *Ammania*, *Polygonum* & *Commelina* etc.

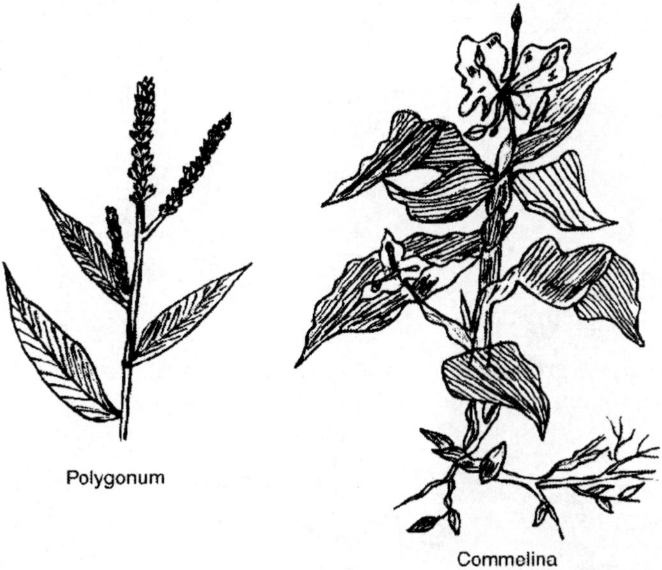

Fig. 4.4: Aquatic Weeds

6. Floating Plants

These weeds have leaves freely floating of the surface of water and roots hanging under earth. They are profusely found in protected ponds and are harmful as they shade the pond example: *Pistia*, *Eichhornia*, *Lemna*, *Azolla, and Salvinia* etc.

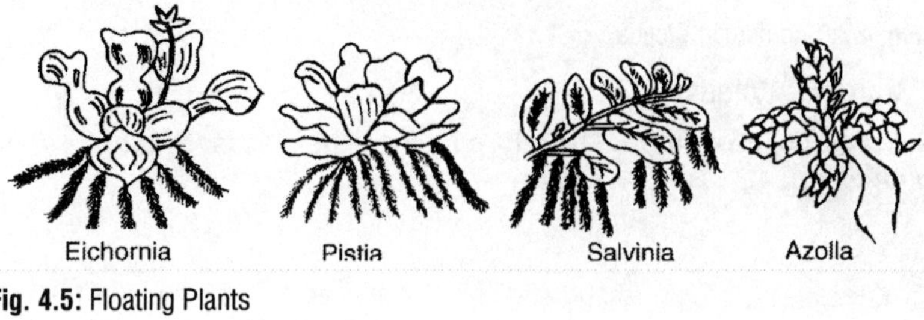

Fig. 4.5: Floating Plants

7. Emergent Weeds

These are surface plants which are rooted in the bottom the pond, but their leaves float on the water surface or rise above the water level. They prefer shallow parts and shores of the pond example *Lemna*, *Nymphaea*, *Nymphoides* etc.

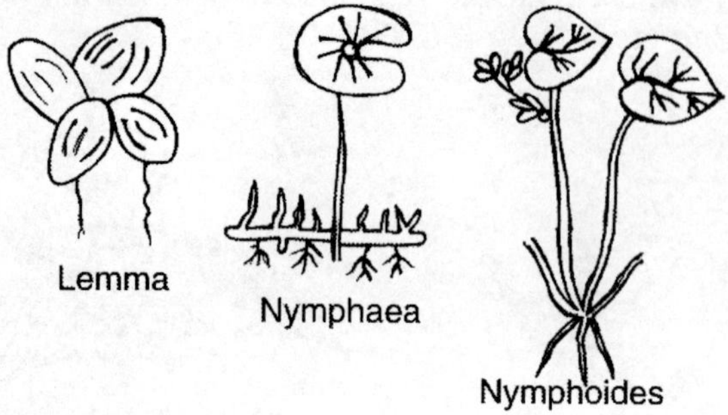

Fig. 4.6: Emergent Weeds

8. Submerged Weeds

These plants grow under water surface and way not be rooted. The rooted submerged weeds are *Najas*, *Vallisneria, and Utricularia* etc.

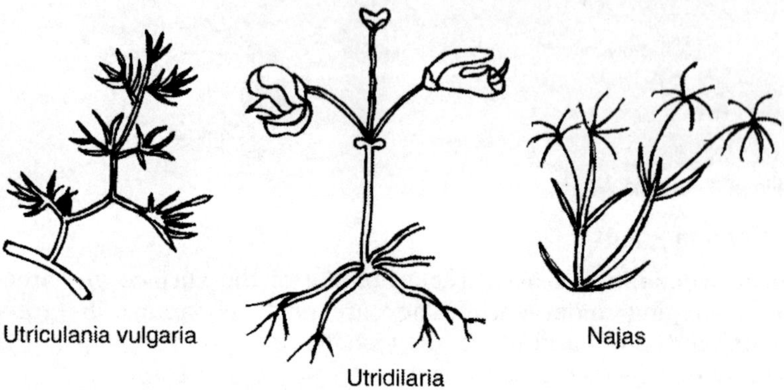

Fig. 4.7: Submerged Weeds

9. Marginal Weeds

These plants grow along the shore lines of the ponds. Example: *Typha*, *Ipomea* etc.

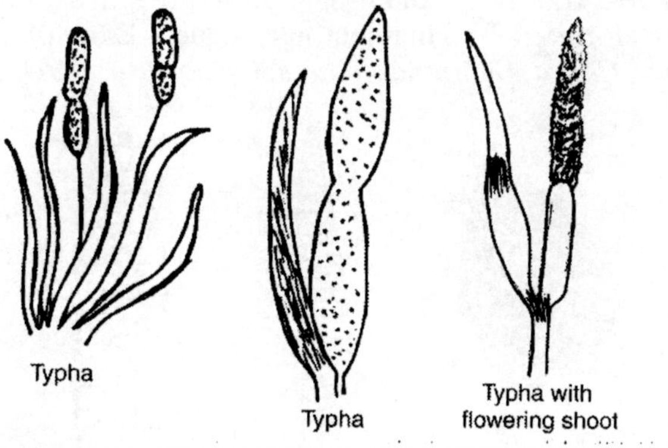

Fig. 4.8: Marginal Weeds

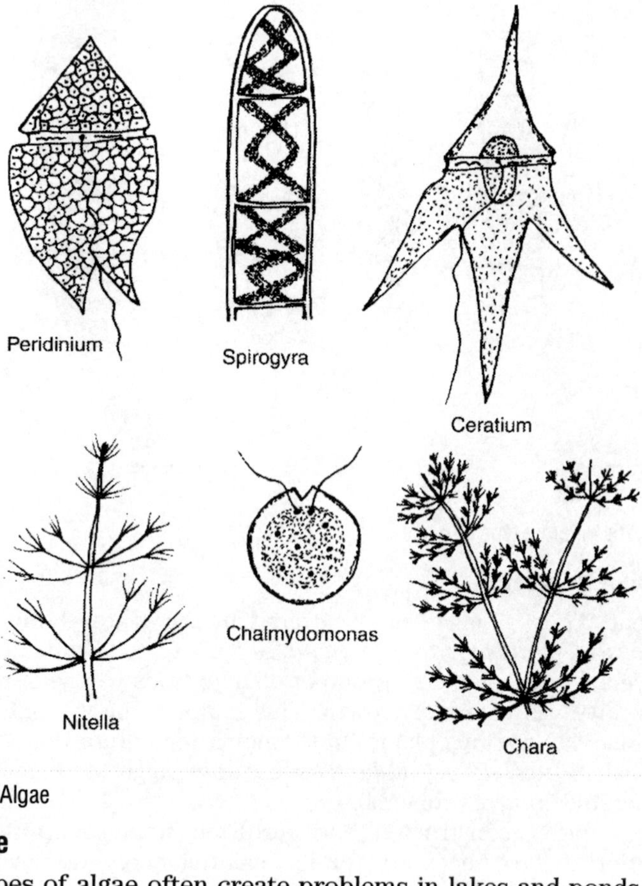

Fig. 4.9: Algae

10. Algae

Many types of algae often create problems in lakes and ponds. These algae may foul the water, and create difficulties in swimming for mats or scums formed by filamentous algae. Examples: *Spirogyra*, *Peridinium Chara*, *Chlamydomonas*, *Cerastium*, *Nitella* etc.

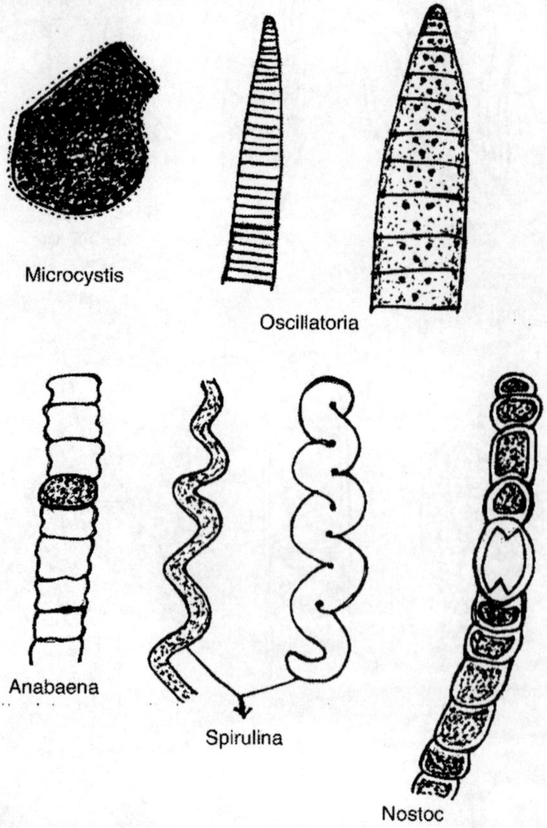

Fig. 4.10: Different type of Algae

11. Different Algae

Different types of algae are scattered in the water body example: *Microcystis*, *Oscillatoria*, *Anabaena*, *Spirulina* etc. Vine climbers and stranglers involve a diverse group of plant species whose one common feature involves vine like growth. This is not to imply that they lack importance as noxious plants. In the north temperate regions species of *vitis*, *smilax*, *Rhus*, *Celastrus* etc. are aggressive and present considerable control problems.

A properly constructed and fertilized pond supports aquatic vegetation that is necessary for its maintenance. But the excessive growth of the plants is harmful and it becomes necessary to control the weeds.

WEED CONTROL

The aquatic weeds can be kept in check by the following methods.

1. Manual and Mechanical Methods

Mechanical removal of weeds by employing manual labor or by machines is the most satisfactory method in India where labour is cheap. Manual labour is often employed. Periodical removal of aquatic plants by hand picking, uprooting or cutting keeps them in check. The rooted submerged weeds are removed by dragging log weeders fitted with spikes and barbed wire, or by cutting them with long handled forks and sickles. During the recent years power lifts or diesel operated machines have been used for eradicating dense rooted vegetation. Several weeds like water hyacinth, Pistia, Salvinia, Azolla and lemna are the most common floating weeds in ponds and lakes. Complete removal of these weeds by manual labour, drying and burning it up. It is considered to be the most effective method of eradicating the weed. The good manure can be obtained by composting the weeds in pits. However, these plants have to be removed before they emerge again.

2. Chemical Controls

Several chemical weedicides are now available for the control of aquatic weeds, but they have to be used carefully to prevent adverse effect on the fishes in water. The chemical is to be selected and used in such a way that-

1. It should be cheap and easily available.
2. Non-toxic to fishes and man.
3. It should not pollute water.
4. It should not involve the use of special and expensive equipments.

Floating Weeds

All floating weeds, *e.g.* the water hyacinth *Eichhornia, Crassipes*, are the most common and of the greatest nuisance value. They can be successfully controlled by the chemical 2,4-dichlorophenoxy acetic acid (2,4-D). This chemical is applied at the rate of 4.5-6.5 kg/hec. and has less harmful effects on fishes. It kills the plants by deadening the leaves and may also be absorbed by the roots destroying them. Pistia and Lemna can also be treated with 2,4-D. Another weedicide is Taficide 80 (2,4-D Sodium Salt 80%) and when used at the rate of 4-6 kg/hec. in an aquosus solution of 1-1.5% concentration in combination with 0.25% solution of "Surf" is reported to kill the water hyacinth completely. Simazine is also an effective weedicide.

7.5-10 lbs of chloroxine in 100-200 gallons of water can be spread on an acre of water surface. This chemical kills the plant in 2-3 weeks but is harmless to fishes and the operator. It has to be used on clear sunny day as rain washes of the chemical from the leaves. Pistia and lemna can also be killed in 10 days by using 1% solution of sodium arsenite (35 Ibs/acre). But this is Poisonous and precautionary measures must be taken. Pistia is also reported to be killed by using ammonia at concentration of 40 ppm.

Marginal Weeds

Plants like Typha Colocasia, Sagittaria and Grasses grow along the pond margin and harbour predatory insects. Although the most effective method is to cut them by manual labour. They can be controlled by chemical weedicides, young Cyperus and colocasia are totally killed with 1-15% aqueous solution of Taficide-80 along with detergent "Surf", colocasia and ipomoea can also be killed with a solution of 2,4-D sodium salt (3-4%) with 0.25%. surf.

Emergent Weeds

Water lilies and lotus are the common rooted plants with leaves emerging on the surface, pulling them out with manual labour or surface treatment with chemicals is of little using eradicating them, use of taficide 80 is reported to be effective in controlling Nelumbo, lily plants get uprooted by treatment with 2,4-D sodium salt at 1.5% concentration, along with 0.25-1% detergent "surf". At the Central Inland Fisheries Research Institute (CIFRI). The marginal weeds like Nymphoides and Nymphaea have been eradicated to some extent by spreading copper sulphate with mud on the bottom soil in 4-5 intermittent doses totaling 175 kg/hec.

Submerged Weeds

Submerged plants like *Hydrilla Vallisneria*, *Potamogeton*, *Najas* and *Ceratophyllum* are the most troublesome and their removal by manual labour is tedious and costly. The submerged weeds are killed if the water is made turbid for a long time. Copper sulphate in combination with ammonia is also found to be effective. Sodium arsenite at 5-6 ppm is also found to be effective in killing submerged weeds, other weedicides like 2,4-D and simazine, when sprayed on water, are only partly effective in killing fully submerged plants.

The inorganic fertilizers like super phosphate and urea have a toxic effects on these plants. Experiments at CIFRI have shown that super phosphate at 500 ppm completely destroyed the submerged weeds. Hydrilla was completely killed with urea at 250-300 ppm. But urea is harmful to fishes as ammonia is released. Anhydrous ammonia is also effective in killing the submerged weeds, but it is also harmful to the fishes.

Algal Blooms and Filamentous Algae

The filamentous algae rarely become a nuisance but sometimes *Spirogyra* and *Phytophthora* form extensive mats. Sodium arsenite is effective in killing them especially in warm weather. Algal blooms consisting of *Microcystis*, *Anabaena* and *Oscillatoria* are sometimes harmful by choking the gills of the fishes by causing oxygen depletion or by rooting. They can be eradicated by treatment with copper sulphate, simazine at a small dose of 0.5-1.0 ppm is also effective in controlling Microcystis.

3. Biological Control

Gupta (1972) has described the biocontrol of aquatic weeds by different bio-agents.

a) **Herbivorous fishes:** Several type of weeds can be kept under control with the help of few species of herbivorous fishes such as the *Ctenopharyngodon idella* (Grass carp) *Cyprinus carpio* (Common carp), *Tilapia mossambica* (Tilapia) and *Osphronemus gourami* (Gourami). The grass carp *C. idella* is an important herbivorous species and effectively keeps certain weeds in check.
It has been found that advanced fry and fingerlings consume *Wolffia*. Advanced fingerlings Juveniles and adults feed on *Lemna*, *Azolla* and *Salvinia*. Submerged weeds like *Hydrilla*, *Najas* and *Ceratophyllum* are consumed by the grass carp. The fish is also reported to feed on *Vallisneria* and *Potamogeton*. In China, the grass carp is reported to consume 40-70% of its own weight of grass per day and 100 fishes per hectare can totally eradicate weeds.
The common carp, *Cyprinus carpio* is also useful in eradicating weeds. While digging and burrowing in search of food, the fish destroys aquatic weeds by uprooting them. The filamentous algae and Pithophora can be kept in check with the help of few carps in the pond. Tilapia and Gourami are also found to be of great help in keeping the submerged weeds under control. Besides fishes, other animals like ducks and geese can also be employed for weed control. Submerged aquatic weeds can be destroyed to a limited extent by shading the pond using surface cover of plants.

b) Manatee: It is a huge water mammal weighting 250 to 1000 kg and has been reported to be a voracious eater of aquatic plants. As it is marine animal, it can be tried in lakes and ponds before it may be used for control of aquatic weeds.

c) **Fresh water snail:** Fresh water snail feeds well on water hyacinth, water lettuce roots and Salvinia leaves. The greatest disadvantages of this bioagent is that it feeds on young paddy seedlings.

d) **The flea beetle:** Flea beetle is a selective aquatic weed eater. It is a specific feeder of alligator weeds.

4. Control of Inorganic Fertilizers

Inorganic fertilizers mixed with 10% Sodium Nitrate are used for controlling submerged weeds like Najas, Potamogeton and Chara. The fertilizers induced heavy growth of phytoplankton and filamentous algae. Which cover the submerged weeds, killing them by decomposition of Urea in 50-250 ppm also bring about decay of submerged vegetation.

Thus, manual labour appears to be a suitable method for marginal and emergent weeds in small bodies of water. A combination of chemical and manual method is more appropriate for controlling hardy aquatic weeds in larger bodies of water. Biological control by using herbivorous fishes is probably the best for controlling weeds. Once the weeds are eradicated proper maintenance of the pond to prevent regeneration and growth is necessary. So, that the pond is best utilized for fish culture.

5. Other Method

a) **Grazing:** A moderate amount of grazing keeps marginal weeds, grasses, shrubs and tree like species under control around ponds. But care should be taken out to allow the cattle to spoil the tanks of ponds.

b) **Dredging:** It is perhaps the commonest method of cleaning weeds from channels and ponds though it is slow process and involves high expenses. Besides it can be used only in open ditches and where at least on side of the ditch is accessible. Dredging may be repeated if there is necessity. This practice removes all types of weeds effectively.

c) **Drying method:** It requires withdrawal of water from the channels during summer for about one week or so. It is quite satisfactorily simple and inexpensive for submerged weeds. To ensure success the channel bottom should be made uniform.

d) **Chaining:** Chaining is generally adopted when there is severe clogging of the ditch. A heavy chain is generally attached between two team of tractors on opposite sides of the ditch. As they move the chain over the weeds break them off. This method is quite inexpensive and is very effective against submerged aquatics. Weeds are completely broken in forward and backward movements whenever, possible, the broken material should be picked out chaining becomes ineffective against emergent weeds.

DISADVANTAGES OF WEEDS

Aquatic weeds are harmful as they consume nutrients of the pond and obstruct netting operations. They check free movement of the fishes and cause oxygen depletion and accumulation of Carbon dioxide, gases like hydrogen sulphite and methane are formed which are harmful to the fishes. Algal blooms choke the gills of the fishes and spoil the water on rotting.

a) Limiting the space for many culturable fishes.
b) Competing with phytoplankton and absorbing nutrients.
c) Causing imbalance in dissolved oxygen concentration.
d) Causing siltation.
e) Hampering, netting operations of harboring unwanted predatory fishes, molluscs etc.

Weeds used for Fish Poisons

i) The bulb of Chlorogalum pomeridianum is used for fish poison.
ii) Leaves of turkey mullein is also used as fish poison.

ADVANTAGE OF AQUATIC WEEDS

Aquatic plants, when in limited quantity, are useful and necessary for the ecology of the pond. They form natural food for many species of fishes, they provide shade and shelter and oxygenate water. They reduce turbidity and provide spawning beds.

ECONOMIC IMPORTANCE

This is a new concept which has originated due to heavy costs entailed on the control of aquatics by manual, mechanical and chemical methods.

Usefulness of Aquatic weeds can be categorized as

1. Weeds as Fertilizer

Compost

Compost is the decomposed plant matter for use as fertilizer in crop fields as well as in fish culture ponds. The compost made from water hyacinth is superior to town compost and farm yard manure. In developing countries where commercial fertilizer is expensive and labour is comparatively cheaper composting the weed to supplement as fertilizer is an ideal proposition.

Besides composting fresh aquatic weed can be used as green manure. Azolla and some blue green algae viz. *Anabaena*, *Nostoc*, *Rivularia* are known to fix atmospheric nitrogen and can be utilized for increasing nitrogen level of fish ponds.

2. Weeds as Feed for Animal/Bird or Fish

1. Weed as Animal Feed

Many types of vegetation that human find inedible can be converted into meat, milk and wool by ruminates like cattle, sheep and goats.

The leafy parts of the aquatic plants such as Dutch weed, water hyacinth and some submerged weeds contain, 25-35% protein, which is exceptionally high. Some submerged aquatic weeds, such as Hydrilla and water milfoil are particularly rich in *carotenes* and Xanthophylls. These valuable pigments are being added to poultry rations in many countries.

In China, pig farmers boil chopped water hyacinth with vegetable wastes, rice, bran, copra cake and salt to make a suitable feed. In Malaysia, the fresh waterhyacinth is cooked with copra meal as feed for pigs, ducks and fish. Aquatic weeds are used as feed by pigs.

Freshly harvested aquatic plants contain enormous quantities of minerals depending on plant type, phosphorus, Magnesium, Copper, Zinc and Manganese were present in same concentrations.

2. Feed for Farm Birds

Ducks, geese, swans and other water fowl feed on vegetation, controlling, weeds on the banks of water ways and often clearing aquatic weeds and algae from small lakes, ponds and canals. Dried duck weeds (30-50 gm) can be added to the daily poultry ration for their weight increase.

3. Feed for Fish

Edward (1980) stated that there are number of herbivorous fishes which directly consume aquatic weeds. Aquatic macrophytes have been known to have potential value as human food, livestock fodder, fertilizers and food for herbivorous fishes. The growth performance food conversion ratio and protein efficiency ratio along with nutritional qualities of weeds what are incorporated in pelleted feeds in fish (Tripathi 1985, Hajra 1987, Das & Ray 1989).

3. Weed as Leaf Protein

Various attempts are being made for the extraction of protein from aquatic weeds. Certain fresh water algae such as Chlorella pyrenoids synthesis has considerable amounts of protein.

4. Weeds as Food Crops

Aquatic plants can provide three types of food, foliage for use as green vegetable, grain or seed that provide protein, starch, oil and fleshy swollen roots that provide carbohydrates.

Some common aquatic plants and their parts are consumed by human.

1. Ipomea aquatica-Young leaves and stem.
2. Nelumbo nucifera-flowers, leaves, rhizome.
3. Euryale ferox-seeds.
4. Trapa-fruit.
5. Colocasia-petiole, rhizome and tubers.

5. Weeds as a Source of Energy

For Present energy crisis, all efforts are being made to find out the unconventional sources of energy from water hyacinth and Salvania.

6. Waste Water Treatment

Many aquatic plants are able to scavenge inorganic and some organic compounds from water eg. Water hyacinth, which is found to be very efficient in this regard.

Ceratophyllum demersum, Spirodela polyrhiza and Lemna minor are used in waste water treatment.

The industrial and domestic effluents containing heavy metals, phenols, phenolic derivatives and pesticide residues are removed to certain extent by water hyacinth. Hoped that water hyacinth could provide safeguard to absorb radio-active isotopes.

7. For Pulp, Paper and Fibre

The common reed Phragmites communist is extensively used in Romania, for making printing paper, cellophanes, card board and various synthetic fibres. The aquatic plants like Typha (Cattails) and Cyperus are the sources of pulp for paper and fibre although these are cultivated as fibre crop.

8. Other uses of Aquatic Weeds

Rough and tough marginal weeds can be used for making huts, thatching roofs, traditional fishing rafts, fishing rods, musical instruments, writing pens and floats. They can be used for making screens and mats.

CONCLUSIONS

It should be emphasized that no method of treatment for weed control in lakes ponds is permanent. Both manual & chemical treatment for the eradication of weeds may have to be repeated twice of thrice a year is found to be effective. Therefore, proper construction of the pond to minimize the probability that weeds will become troublesome and proper utilization for best long period, weed control measures that can be recommended.

Table 4.1: Eradication of aquatic weeds by chemicals

Type of weed	Chemical and dose	Method of application	Remarks
Blue-green algae and algal weeds.	a) 1ppm copper sulphate. b) 0.5-1.0ppm Simazine. 0.1-0.3ppm Diuron (1kg/ha) 0.2ppm Dequat (1kg/ha) 0.2ppm Paraquat (1kg/ha)	Copper sulphate packed I cloth bags is immersed in water; applied once in week if needed for a number of weeks; other chemicals are sprayed	Copper Sulphate is economical, easy for use and safe; safe level for human 2.5ppm and for cattle 100ppm.
Floating weeds.	a) 80% 2-4 dichlorophenoxy acetic acid (5-7 kg/ha) b) 80% 2-4 D (4-6kg/ha+small quantity of 0.1% surf detergent. c) Simazine (5kg/ha) d) 15 ppm ammonia. 0.2ppm Dequat (1kg/ha) 0.2ppm Paraquat (1kg/ha)	Spraying.	----------
Emerged weeds.	a) 80% 2-4D (4-6kg/ha)+Small quantity of 0.1% surf detergent. b) 5% Dalapon (2-2 dichloropropionic acid) (5-10 kg/ha) c) Dichlorophenyl (2-6 dichlorobenzonitrile) (10kg/ha) d) Fenae (2,3,6 trichlorophenyl acetic acid) (10 kg/ha)	Spraying.	Dalapon solution is made by diesel oil; dichlorophenyl and Fenae applied after draining the water.
Submerged weeds.	a) 15ppm ammonia b) 3-5ppm Simazine. c) 250ppm urea. d) 4ppm sodium arsenite. e) 500ppm Superphosphate. f) 40ppm Sulphur dioxide.	Spraying	(a) Sodium arsenite is to be carefully handled as even a dose of less than 0.05 is unsafe for human drinking, however, fish tolerate upto 12 ppm. b) Endothal is not widely used as even doses less than 0.3ppm are toxic to fish.

Type of weed	Chemical and dose	Method of application	Remarks
	0.2ppm Dequat (1kg/ha) 0.2ppm Paraquat (1kg/ha) Dichlorophenyl (2-6 dichlorobenzonitrile) (10 kg/ha) Fenae (2,3,6 trichlorophenyl acetic acid) (10kg/ha)		c) Dichlorophenyl and Fenae are to be used after draining the water from fish pond.
Marginal weeds	a) 80% 2-4 D (5kg/ha) +Small quantity of 0.1% surf detergent b) 0.2ppm Dequat (1ka/ha) c) 0.2ppm kParaquat (1kg/ha) d) Dichlorophenyl (2-6 dichlorobenzonitrile) (10kg/ha) e) Fenae (2,3,6 trichlorophenyl acetic acid) (10kg/ha)	Spraying	Dichlorophenyl and Fenae are to be used after draining the water from fish pond.

Chapter 5

Larvivorous Fishes

INTRODUCTION

Larvivorous Fishes feed on larvae of different small animals (Insects) and was discovered by Alkins in (1901), Chatterjee in (1939), Hora in (1954) and Prasad in (1956). These fishes are used in controlling aquatic organisms and Larval forms of mosquitoes. Mosquitoes are the vectors of several diseases and carry parasites of Malaria, Filaria, Yellow fever etc. They breed in all sorts of stagnant water and shallow weed infested ponds, swamps, pits, gutters, and all other kinds of inland water bodies. Different species of mosquitoes prefer different kind of water for breeding purposes. During the recent years, DDT and other insecticides are being used in increasing quantities to control mosquitoes, but these chemicals have their effects on human health also. Fish is a natural enemy of mosquito eggs and larvae and its use as a means of biological control has been recognized since olden times. The larvivorous fishes consume the mosquitoes, larvae, adult mosquito, crustacean, Cyclops and molluscan larvae. Here by preventing their related disease some species of fresh water fishes are known to be larvicidal and used in the control as mosquitoes and other larvae.

DEFINITION

These species of fishes prefer mosquito larvae and feed on them throughout their life are known as larvicidal fishes.

Special Features as Larvicidal Fishes

1. The larvicidal fishes should be of small size so as to be capable of moving freely among weeds in shallow waters.
2. It should be a hardly fish, surviving in deep as well as in shallow water and with-stand transportation to long distances.
3. The larvae normally occur at water surface and remain attached to submersed aquatic vegetation. The larvicidal fishes should be surface feeders and will be able to spot and eat the larvae.
4. The larvicidal fishes should breed freely in confined water.
5. The fishes must be fast swimmers.
6. The larvicidal fishes should be carnívoros in habit.
7. The fins should be spineless.
8. Pelvic fins should be abdominal.

9. The fishes must be able to adopt to certain adverse conditions in water like, change in pH, temperature and various concentration of pollutants.
10. These fishes should be variously coloured.

TYPE OF LARVIVORUS FISHES

Exotic Fishes

1. Carassius auratus

It is commonly known as gold fish and has been imported from various countries, mainly as an ornamental fish and can be used for mosquito control. But due to its high cost, its use on a large scale is not possible and its utility in stagnant water is found to be less useful (Fig. 5.1)

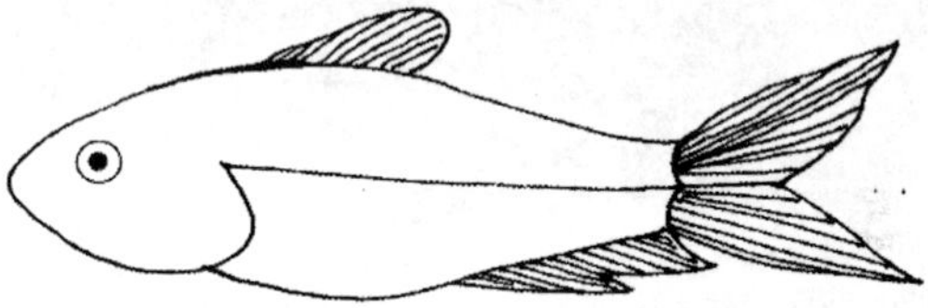

Fig 5.1: *Carassius auratus*

2. Lebistes

Lebistes has also been brought to India from America. It is found in the ponds and tanks in South India. It is a viviparous fish and can also tolerate polluted condition of water. (Fig. 5.2)

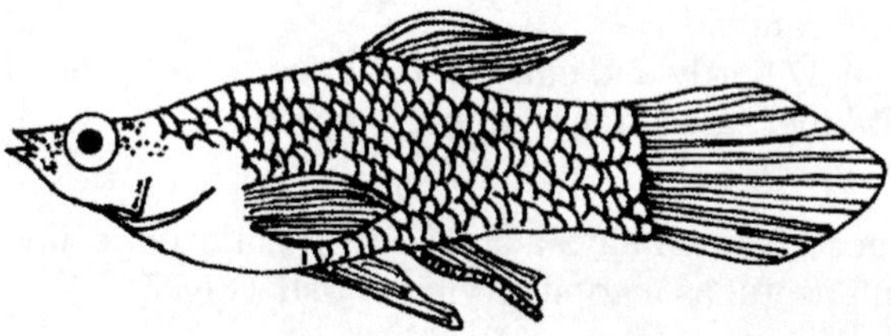

Fig 5.2: *Lebistes reticulatus* (Male)

3. Gambusia affinis

Gambusia the famous fish known as 'Topminnow' has been extensively used for mosquito control. It was brought to India in 1931 from Italy. It is very voracious and destructive fish. It is about 2½" in size. It starts breeding when it is only 2" in size. It can with-stand all types of foul water. It is a predaceous fish and very active for its size.

It possesses a great liking for the larvae of mosquitoes. It is also known as mosquito fish. Black spots are observed on the body. (Fig. 5.3)

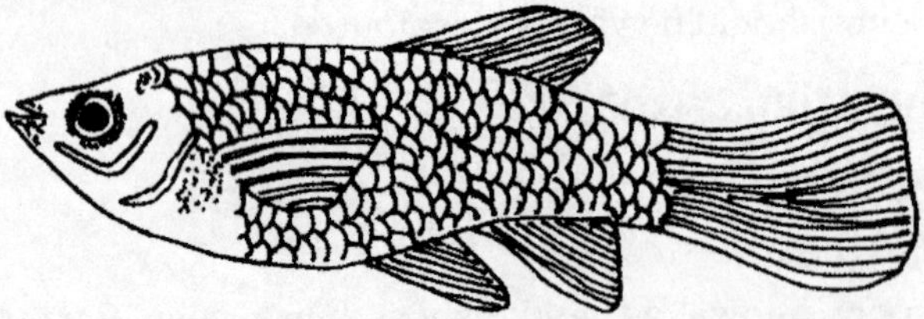

Fig 5.3: *Gambusia affinis* (Female)

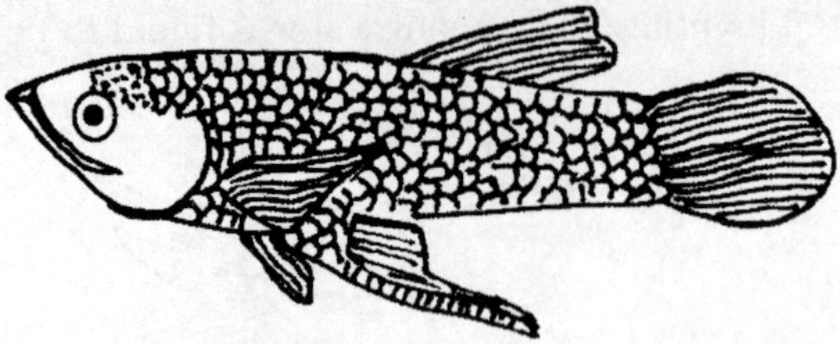

Fig 5.3a: *Gambusia affinis* (Male)

4. Gambusia holbrooki

It is named in the honor of the naturalist J. E. Holbrooki. It is about 3" in size. Females normally breed from May to October. The fish has been introduced from California, Philippines and Japan. These fishes have proved to be a successful destroyer of mosquito larvae. One individual fish is capable of eating 165 larvae in 12 hours as recorded by S. F. Hilderebrand.

It is extremely hardy and can live in all winter months under a thin sheet of ice. It hibernates in mud during winter.

5. Tilapia

It was introduced from Malaysia and Mauritius for experimental work. It is now found useful as a good larvicidal fish. (Fig. 5.3b)

Fig 5.3b: *Tilapia mossambicus*

6. G. nicaraguensis

It has been names after the lake Nicaragua. It is about 2½" in size. Temperature requirement of this fish is 50 to 90 °F. It is a successful mosquito fish. The colour is grayish olive with a metallic blue shade on the sides. It has scattered brown spots on all unpaired fins and body.

Indigenous Fishes

There are several indigenous species same of which are very effective in destroying mosquito larvae.

1. Notopterus notopterus

This is a fish that serves both as larvicidal as well as a food fish. It is considered to be useful for mosquito control. The young stages of this fish fed on mosquito larvae but the utility of the adult is very doubtful. The adults feed on small fishes and is therefore harmful (Fig. 5.4)

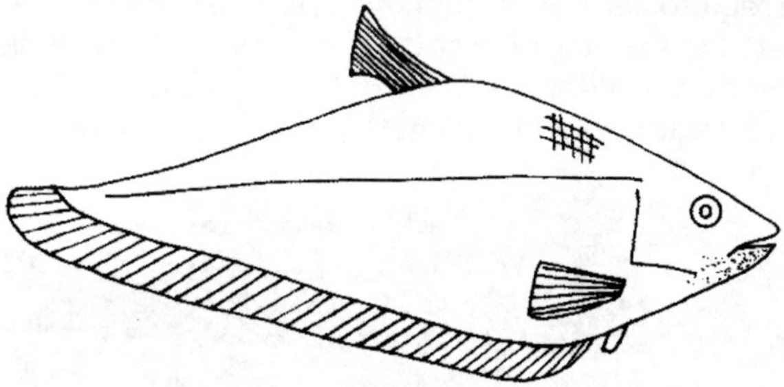

Fig 5.4: *Notopterus notopterus*

2. Laubuca laubuca

Laubuca is represented by two species in India which are considered as larvicidal forms. They are delicate fishes and not able to withstand transport to long distances. (Fig. 5.5)

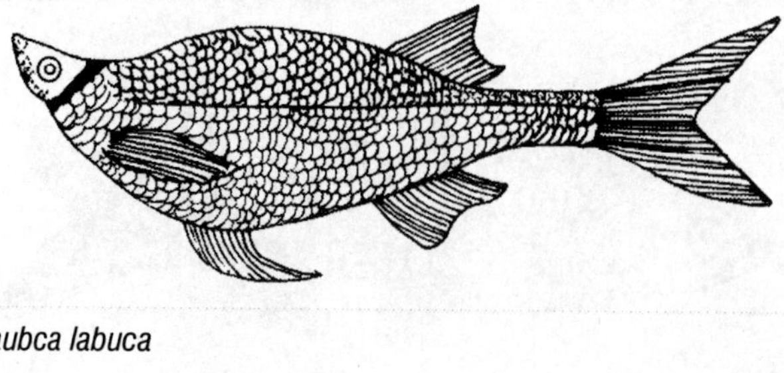

Fig 5.5: *Laubca labuca*

3. Barilius

Barilius is found chiefly in the hill streams. The two species (*B. ben*

Del Isis, *B. Vagra*) which are small in size are considered effective larvicidal forms. (Fig. 5.5b)

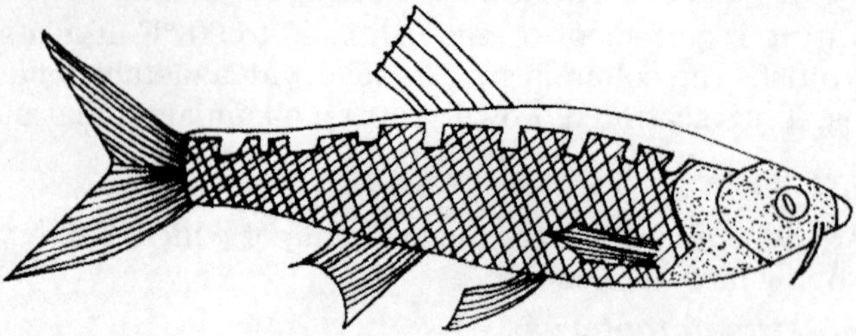

Fig. 5.5b: *Barilius vagra*

4. Oxygaster

It is commonly known as "Chilwa or Chilara". The smaller varieties feed on mosquito larvae throughout life. They have upturned mouth and feed at the surface of water. They are delicate fishes and their larger species are used as food. There are several species of oxygastor which are of great utility in mosquito control. (Fig. 5.6)

Fig. 5.6: *Oxygaster bacaila*

5. Rasbora daniconius

Rasbora daniconius is the most widely distributed species and regarded as an effective larvicidal form, but is not suitable for extensive use. It can be of use in regions where it occurs abundantly. (Fig. 5.7).

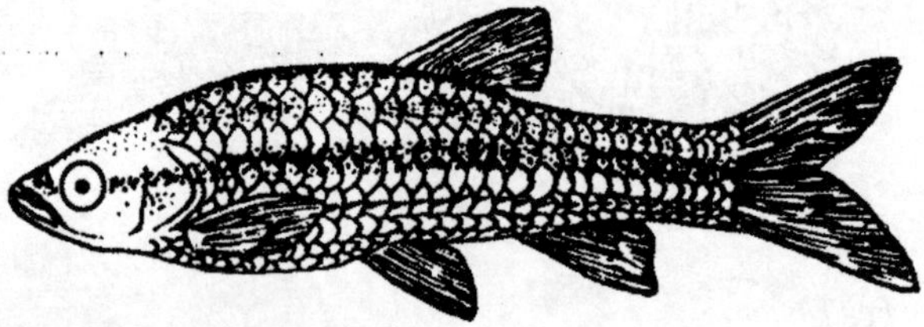

Fig 5.7: *Rasbora daniconius*

6. Esomus danricus

Esomus danricus was reported by Choudhary in (1911) and Hogdson in 1914. It is commonly called flying bards. Esomus is a small size fish used in aquaria, where it breeds feely. It is of great utility and can be employed on large scale for malaria control. (Fig. 5.8)

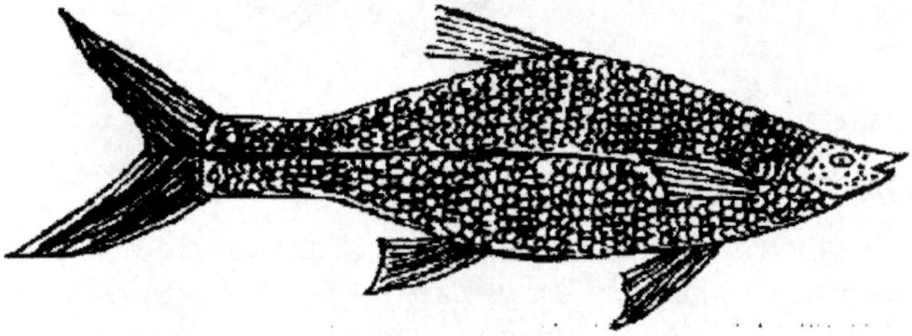

Fig. 5.8: *Esomus danricus*

7. Danio rerio

It is commonly found in plains. It is reported to be a good larvicidal fish and regarded of some value in mosquito control (anti malaria) (Fig. 5.9).

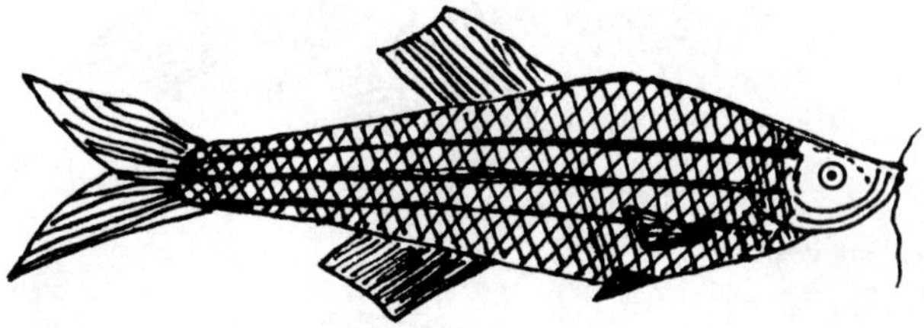

Fig. 5.9: *Danio Rerio*

8. Aphanius dispar

Aphanius dispar is the only species which occurs in India, which has been found to be an effective destroyer of mosquito larvae (Fig. 5.10)

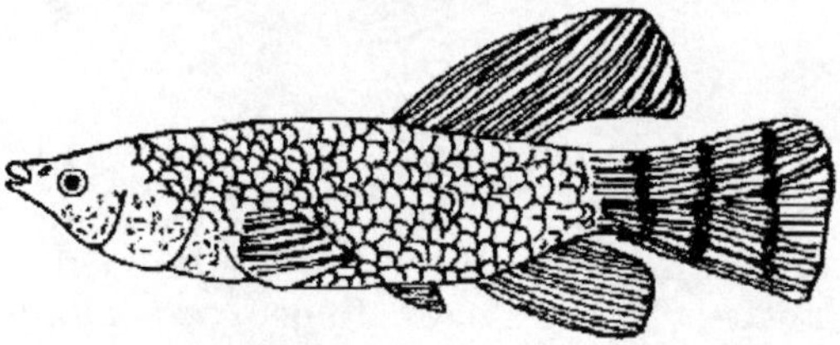

Fig 5.10: *Aphanius dispar* (Male)

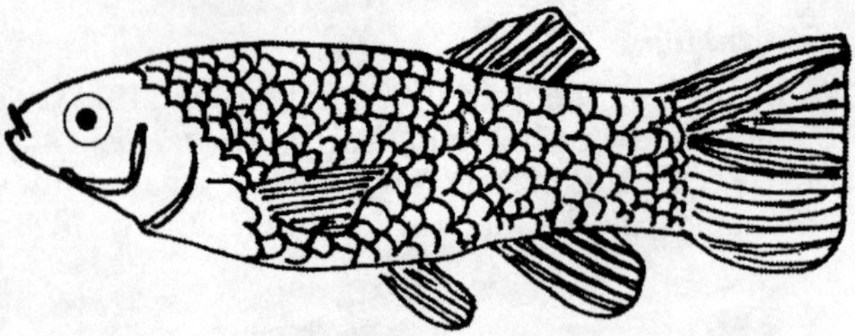

Fig. 5.10a: *Aphanius dispar* (Female)

9. Puntius

There are several species of Puntius and they are popularly called "Carp minnows". They are small hardy species and practically have no value as human food. They can withstand transport and breed in confined waters. They are useful as larvicidal fish. (Fig. 5.11).

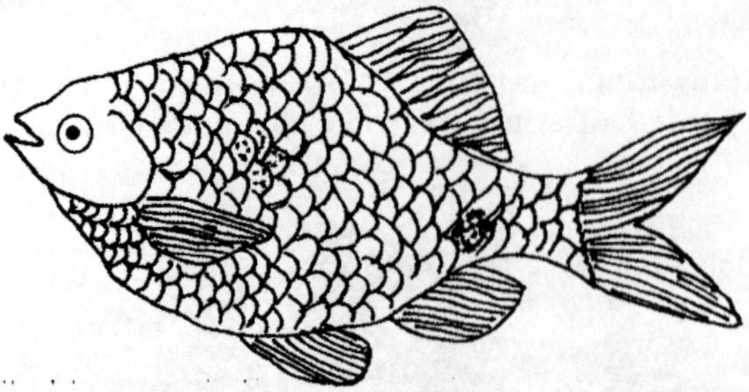

Fig. 5.11: *Puntius ticto*

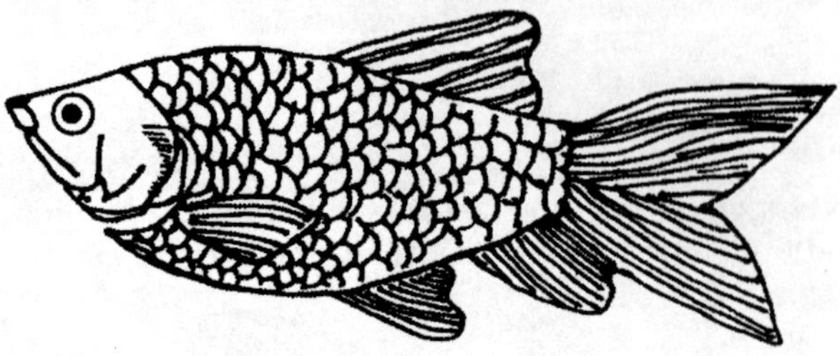

Fig 5.11a: *Puntius sophore*

10. Carps

All major carps Cirrhina, Labeo and Catla feed on insects and crustaceans in their young stages. Hence carp culture is useful for mosquito control as well as for providing food fishes, but cannot be called as larvicidal fishes.

11. Aplocheilus panchax

Aplocheilus panchax known as "top minnows". This species is found throughout India. It is a hardy fish and can live in all sorts of water. It is most important indigenous larvivorous species. (Fig. 5.12)

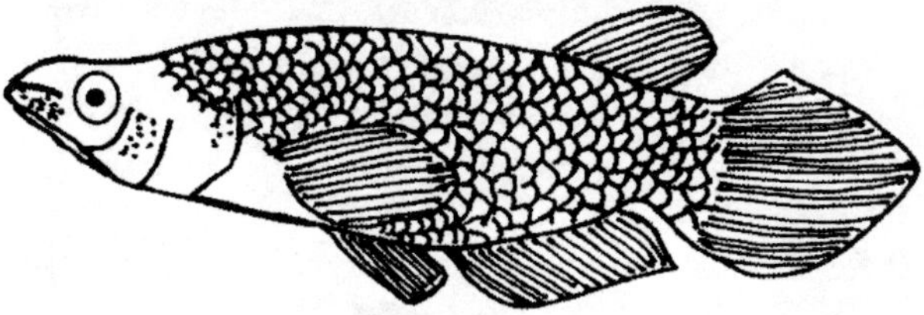

Fig. 5.12: *Aplocheilus panchax*

Fig. 5.12a: *Aplocheilus lineatus*

12. Oryzias melanostigma

Oryzias is closely allied to *Aplocheilus*. It is very useful as larvicidal form but cannot be easily transported as it is a delicate fish. (Fig. 5.13).

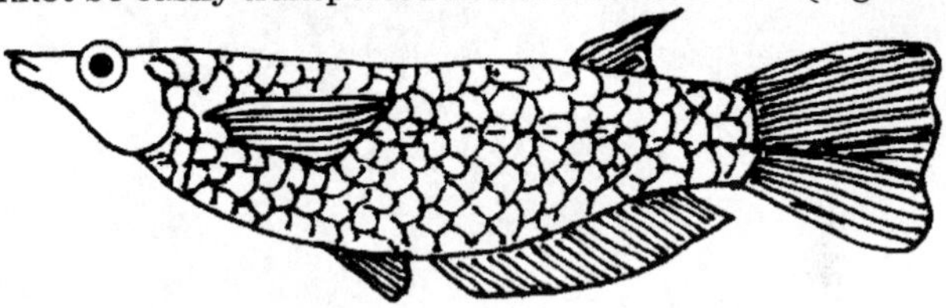

Fig. 5.13: *Oryzias megastigma*

Colias fasciatus

It is a small beautifully coloured, hardy fish and is widely distributed. It breeds in confined waters. It is regarded to be a good larvicidal fish and serves as food fish in Bengal. (Fig. 5.14).

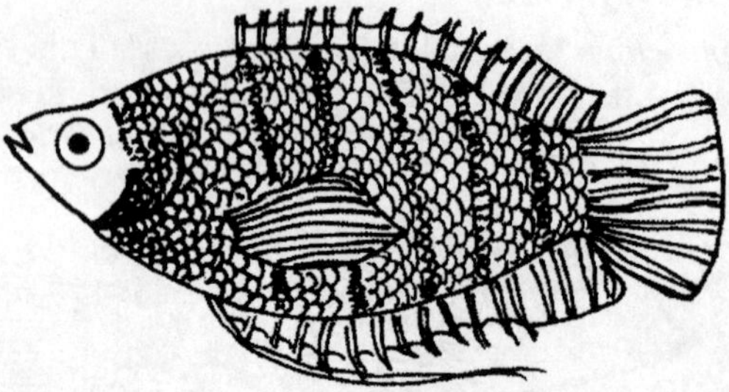

Fig. 5.14: *Colias fasciatus*

Macropodus cupanus

M. Cupanus is found in Malabar Coast. It is a small hardy fish. It breeds freely in ditches, paddy fields and shallow waters. It is considered as a valuable carnivorous fish.

13. Etroplus

Etroplus represented by two species in India. *E. suratensis* is and *E. maculates*. It lives in fresh water as well as in brackish water. The adult feeds on algae but young ones feed on larvae. (Fig. 5.15)

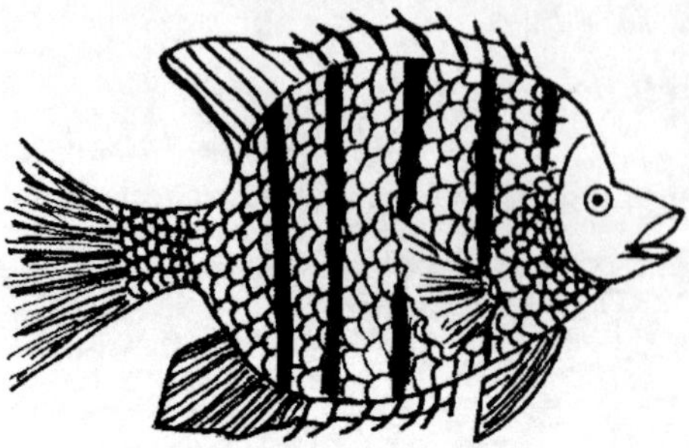

Fig. 5.15: *Etroplus suratensis*

14. Anabas

It is represented by only one species *Anabas testudineus* and commonly called as "climbing perch" It is hardy fish due to the presence of accessory breathing organs. It is reported to be a useful larvicidal fish and is of great commercial importance. (Fig. 5.16).

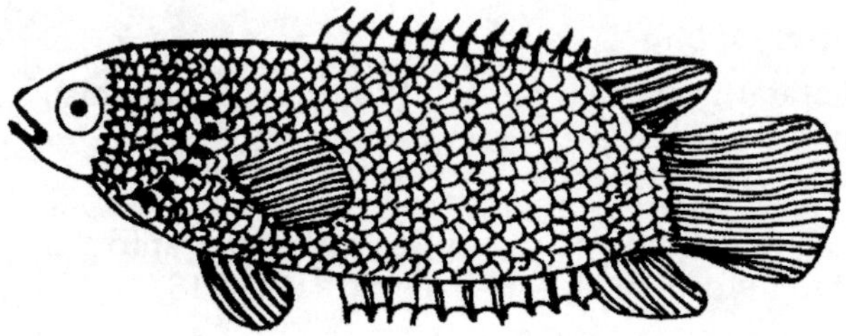

Fig. 5.16a: *Anabas*

15. Mugil

They are commonly known as mullets. Some species live in estuarine and fresh water. The young ones are surface feeders and destroy mosquito larvae but it cannot be used for extensive malaria operations. (Fig. 5.16b)

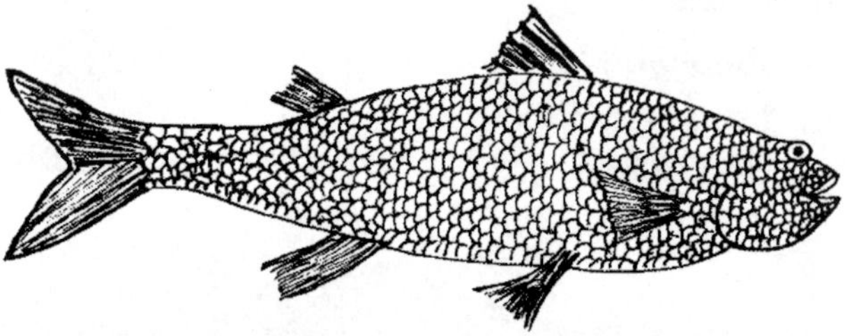

Fig 5.16b: Indigenous *Lindorus* fishes

16. Ambassis

Two species *A. nama* and *A. ranga* are widely distributed. They are of small size and translucent fishes. They feed mainly on Cyclops & crustaceans, but can feed eggs and larvae of mosquito. They can be used for mosquito control also. (Fig. 5.17)

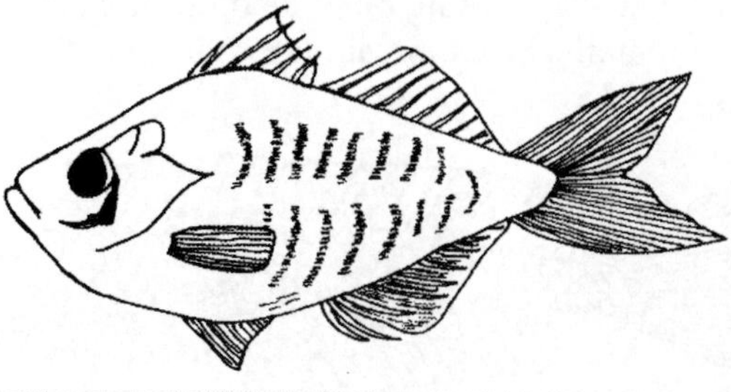

Fig. 5.17: *Ambassis*

17. Terapon jarbua

It ascends tidal water and is useful for destroying mosquito larvae in brackish water.

18. Badis

It is a small beautifully coloured fish and grow upto 2-3 inches. They are reported to feed on mosquito larvae. (Fig. 5.18).

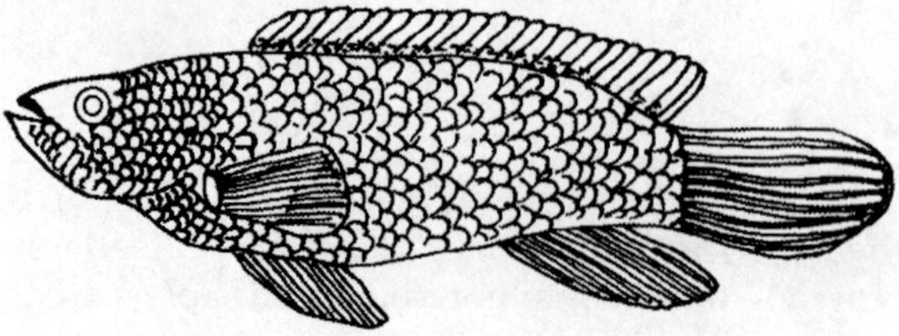

Fig 5.18a: *Badis badis*

19. Glossogobius giuris

It is widely distributed carnívoros fish. In young stages, it feeds on crustaceans, insect eggs and larvae of mosquito. (Fig 5.18b)

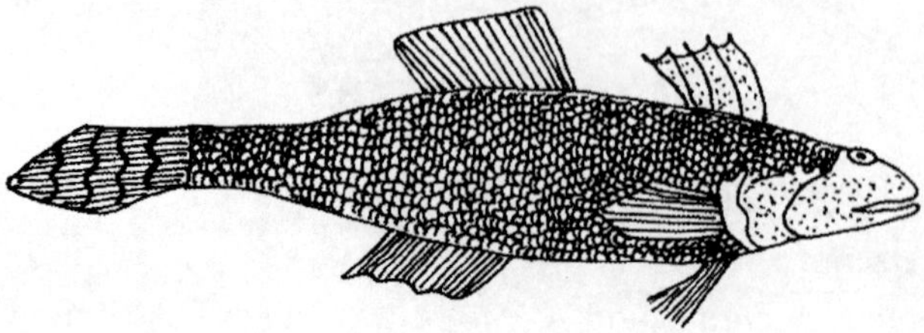

Fig. 5.18b: *Glossogobius giuris*

20. Channa

It is a predatory fish. The young ones feed on mosquito larvae but the adults feed on small fishes. It is not suitable for anti mosquito operation. (Fig. 5.19)

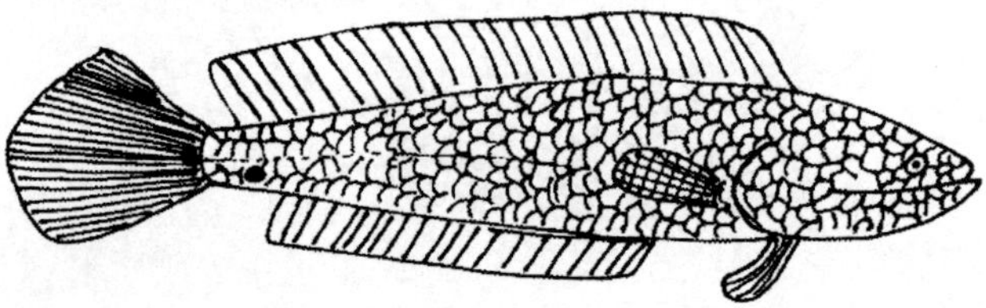

Fig. 5.19: *Channa striatus*

RATES OF CONSUMPTION OF LARVAE

All larvivorous fishes consume the larvae at different rates. The list given below showing the rates of various larvivorous fishes Table 5.1.

EFFICIENCY OF LARVICIDAL FISHES

Job (1953) divided the mosquitocidal activity of these fishes into six groups, according to their efficiency.

1. *Aplocheilus and Gambusia* are typical surface feeders which are primarily mosquito fish and highly efficient for anti-malaria work.
2. *Oryzias*, *Lebistes*, *Athanius* and *Horaichthys* are surface feeder but less efficient due to their mode of life.
3. *Damion*, *Rasbora*, *Esomus* and *Craniums* are sub surface feeders and are larvicidal to a considerable extent.
4. *Puntius*, *Ambassis*, *Anabas*, *Bodis* are column feeders which feed on larvae when they get a chance.
5. Large size food fishes are labeo, catla, mugil etc. whose fry feed on larvae.
6. *Wallago*, *Channa*, *Notopterus* and *Mystus* are predatory fishes, whose fry feed on larvae, but adult fishes are destructive for other fishes.

Table 5.1:

S. No.	Name of fish	No. of Larvae eaten
1.	Rasbora	400
2.	Gumbusia	375
3.	Oryzias	300
4.	Puntius	250
5.	Channa ranga	190
6.	Puntius ticto	180
7.	Laubuca	170
8.	Gambusia Holbrooki	165
9.	Barilius vagra	165

Hora and Mukherjee (1953) have classified the larvivorous or larvicidal fishes into the following groups

1. Typical Surface Feeders

Gambusia, *Aplocheilus* etc. are mosquito fish and are efficient for anti-malaria work.

2. Surface Feeders

Surface feeders but less efficient to their mode of life Oryzias,

Lebistes, Aphanius etc.

3. Sub Surface Feeders

They are larvicidal to a great extent *e.g.* Danio, Rasbora Esomus and Carassius etc.

4. Column Feeders

They feed on the larvae of mosquitoes when they get chance *e.g. Puntius*, *Badis*, *Colisa Anabas* and *Macropodus* etc.

5. Carps

The fry of carps feed on larvae *e.g. labeo*, *catla*, *mugil* etc. *catla*, *mugil* etc.

6. Predatory Fishes

The fry of *Wallago*, *Channa*, *Notopterus* and *Mystus* feed on larvae but the young and adult fishes are not suitable for mosquito control. From the above account it is clear that these fishes belongs to the family Cyprinodontidae, which feed on mosquito larvae.

DRAW BACKS OF LARVICIDAL FISHES

i) When the water is highly polluted, the larvae increases, but the larvicidal fishes become sluggish and therefore, feeding is decreased.
ii) During rainy season, water becomes turbid and larvae find their way without being caught by the larvicidal fishes.
iii) High changes in water conditions may cause death of fishes.
iv) If the water become deeper, the larvicidal fishes have to face the danger of being eaten up by other aquatic large predators.

IMPORTANT LARVIVOROUS FISHES

A large number of species of fishes, both indigenous and exotic and larvivorous in nature, some fishes have been listed below.

1.	Gambusia affinis (a cichlid)	An exotic fish.
2.	Lebistes reticulates (a cichlid)	An exotic fish.
3.	Cranius earamium. (Cyprinid fish, gold fish)	An exotic fish.
4.	Aplocheilus lineatum (A minnow)	An Indigenous
5.	Oryzias melanostigma (Minnow)	An Indigenous
6.	Aphanizius diaper (A minnow)	An Indigenous
7.	Amblypharyngodon (A cyprinid)	An Indigenous
8.	Danio rerio. (A cyprinid)	An Indigenous
9.	Rasbora daniconius (A cyprinid)	An Indigenous

10.	Esomus danricus (A cyprinid)	An Indigenous
11.	Puntius sophore (do)	An Indigenous
12.	Colisa fasciata (Anabantidae)	An Indigenous
13.	Barilius Vagra (a cyprinid)	An Indigenous
14.	Notopterus notopterus (An Osteoglossidae)	An Indigenous
15.	Oxygaster (A cyprinid)	An Indigenous

CONCLUSION

The actual control work has to be planned carefully considering the ecological features of the area to be treated. A suitable species can be introduced directly into the water after clearing dense surface vegetation that might be present. Any predaceous fish present in the water should also be removed before introducing larvivorous fishes.

Chapter 6

Hill Stream Fishes

INTRODUCTION

The fast flowing hill streams, rivers and rivulets call for a special modification of the fish fauna. The individual modification is chiefly manifested in being small sized development of structures for adhesion. A good number of fishes, migrate from sluggish waters of the lower streams, reaches to colonize in the torrential water of upper streams. These fishes migrate mainly in search of food, shelter and spawning.

The hill stream fishes may be divided into two groups. One group is temporary inhabitants of the hill streams and migrate up only at certain periods of their life for specific purpose of spawning.

Second group live permanently in the rivers and streams of the hills. Several structural modifications have been seen in these fishes. There is a great range of variation of characters in the hill stream fishes. Some important contributions on the various aspects of hill stream fishes are given by numerous scientist.

Chaudhari (1909), Jenkins (1910), Hora (1921,22,23), Menon (1954,1964), Haws and Goodnight (1961), Menon and Dutta (1964), Lal *et al.* (1966), Ganguly *et al.* (1972), Tilak and Hussain (1974), Joshi and Jordon (1977), Bhatt and Singh (1981), Shah & Agarwal (1985), Pandit and Yousuf (1984), Singh et al. (1989), Hussain (1989) and many others.

ORIGIN OF HILL STREAM FISHES

There are two different views regarding the origin of hill stream fishes. According to the first view, the fishes inhabiting the hill stream reached there accidentally.

The second view suggest that there has been step by step colonization of the mountain streams from the sluggish streams of the plains and fishes, migrated to upstream in search of food and protection from the predators which later becomes specialized to new condition.

The first possibility appears to be improbable if the fishes reached up streams and they have not any adaptive feature they would not have been able to survive in rapid streams. The second view suggested by Hora (1922) appears to be more correct. The following facts support the second view.

1. The fishes marked from the sluggish water at the foot hill to the rapidly flowing waters on mountain hills eg. *Crossocheilus latius*.
2. The migrated fishes become gradually specialized to the condition of hill streams ex. *Garra* and *Glyptothorax*.
3. A study of the development of the form of the body, air bladder, position of mouth, eyes etc. suggest their gradual evolution from such forms that lived in the sluggish water; example *Garra*.
4. In many instances only a few members as a family and in many others all the members of a family became adapted to conditions of hill stream. In Loricariid catfishes. The lower lands forms are well armoured while their hill stream forms are naked.
5. The *Crossocheilus latius* appears to be an intermediate form to live both in sluggish and rapid water of its homeland.

IMPORTANT HILL STREAM FISHES

The important hill stream genera belonging to three distinct families are given below:

Balitera, *Bhavania*, *Psilorhynchus*, *Garra*, *Psilorhynchus*, *Schizothorax* , *Nemacheilus* , *Barilius*, *Barbus*, *Crossocheilus* are all fishes belongs, to order cypriniformes and family cyprinidae. *Glypto Sternum Glyptothorax*, *Pseudoephedrine laguria* and *erethistes* belong to order siluriformes and family sisoridae, and *Nemacheilus botia* belong to family Cobitidae. Mishra (1984) reported some species of hill stream fishes from Nepal and Bihar. Some important among them are-

1. Labeo dyocheilus, (Meclelland)
2. Labeo microphthalmus (Day)
3. Puntius terio (Ham.) Garra Gotyla, Stenorhynchus (Jordan) Botiageto (Ham.) Lepidocephalus annandalei (Choudhari)
4. Nemacheilus, rupicola (Mcclellandi).

PHYSICAL CHARACTERISTICS OF HILL STREAM

The conditions which are very adverse influence the fish, living in hill streams for specialization.

Strength of Water Current

The chief factor is the strength of the current of water which influence the evolution of fish fauna. The rate of flow of water varies in different seasons. It is in general, much higher than in rivers and streams of the plain. Various obstacles in the course of the water current, however, determines the laminar turbulent of shooting flow of water. The fishes, have therefore, to develop adhesive organs to withstand the rapid flow of water currents.

Availability of Food

Sufficient amount of food is available in the hill streams. The fishes have to depend upon the periphyton constituting the primary key to productivity. The vegetation cannot grow due to rapid flow of water. Therefore they largely depend open the algal filaments forming slime covering on the rocks. In certain regions the microorganism and the insect larvae may also become available to them.

Intensity of Light

The sun rays in hill streams penetrate deep into the water because the water is shallow and very clear owing to the absence of suspended particles. The fishes have to bear intense light during the day or shelter themselves under the stones or rocks.

Dissolved Oxygen

The water in mountain streams is under constant motion, and it is well aerated. Plenty of oxygen is therefore, a favourable condition to fishes.

Temperature

The temperature of hill streams fluctuates rapidly but remain more or less constant from surface to the bottom. The water is generally cooler but gets heated in the sun.

MODIFICATION OF HILLS STREAM FISHES

Size

Hill stream fishes are generally of small size, they have streamlined body and large heads, so that they can hide under the rocks stones. There are, however, two exception belong to family Catostomidae, but which have adapted secondarily to mountain streams.

Shape

These fishes are generally cylindrical and dorsoventrally flattened (eg. Balitora, Glyptosternon and Garra), with the ventral side straight and the dorsal slightly arched. These fishes press against the bottom with their flattened ventral surfaces and so prevent themselves from being swept away with the water currents. The head is generally small and pointed. (Fig. 6.1)

Scales

The scales are generally small and partly buried in the skin. It is because of this fact that they are free from the danger of predators. The scales are absent in scapular and anal region. The ventral scales are generally absent and used for attachment to the rocky bottom. No satisfactory reason is known as to why the scales are absent. It is only claimed that it provides the resistance to the fish in water current. (Fig. 6.2)

Caudal Fin and its Peduncle

Hill stream fishes have long, narrow band shaped caudal fin which is supposed to be an adaptation for fast flowing water, and high altitude example Nemacheilus: The lobes of caudal fin are generally unequal, the lower being longer than the upper *e.g.* in *Garra*, *Balitora* etc.

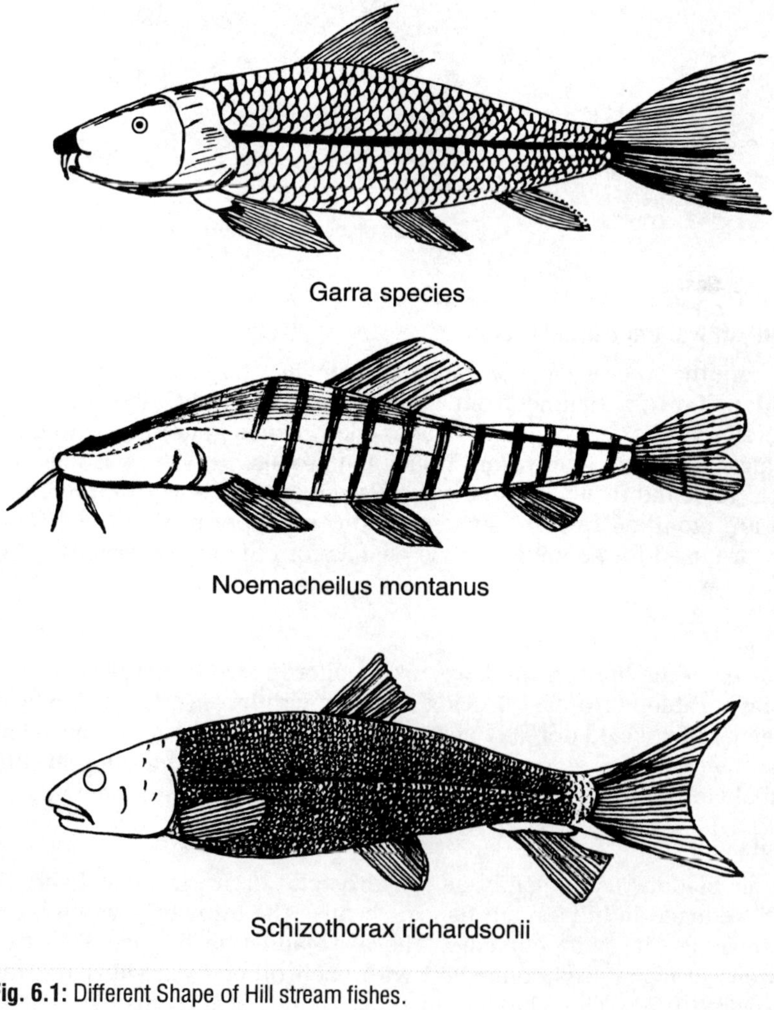

Fig. 6.1: Different Shape of Hill stream fishes.

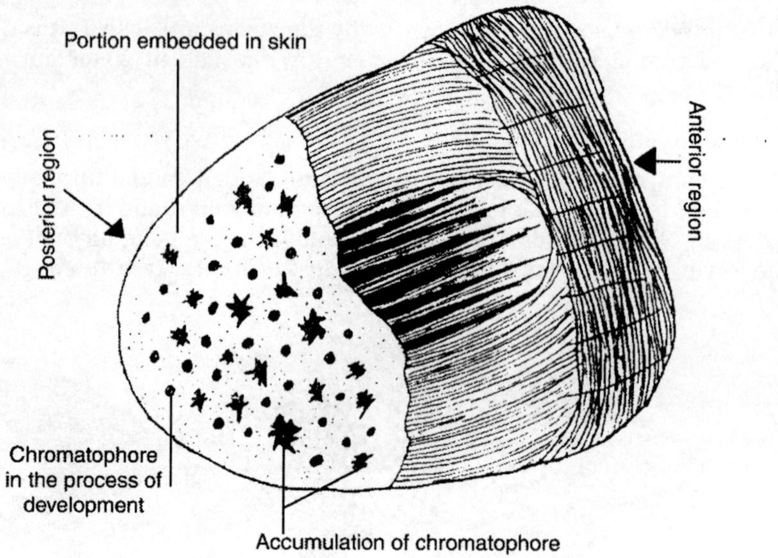

Fig. 6.2: Scales

Mouth, Jaws and Barbels, Lips

The mouth is crescent or semi-circular in shape and ventral in position, fairly a behind from anterior end of snout. The jaws are covered with horny covering because these fishes generally scarab off the algal food and slime from rocks and stones etc. The barbells are short, thick and fleshy at the base in *Pseudecheneis* and *Glyptothorax*, Lips are modified to form suckers of diverse type in different fishes. They are used for scoping mud as well as for clinging on stones. (Fig. 6.3)

Eyes

The eyes of the hill stream fishes are smaller in size, as larger eyes are not favourable in the fast flowing streams with intense light. The eyes are generally placed dorsally near each other. The eyes are regarded as protective measure against the intense sunlight, while their dorsal shift is an obvious adaptation, to free the ventral surface for attachment.

Air Bladder

The air bladder used chiefly as a hydrostatic organ in most fishes is much reduced in hill stream fishes, because the buoyancy would be a disadvantage in swift currents. The air bladder is divided into two portions more or less connected with each other by a short narrow transverse tube. The bladder, in most of hill stream fishes remain enclosed in a thin bony capsule. (Fig. 6.4)

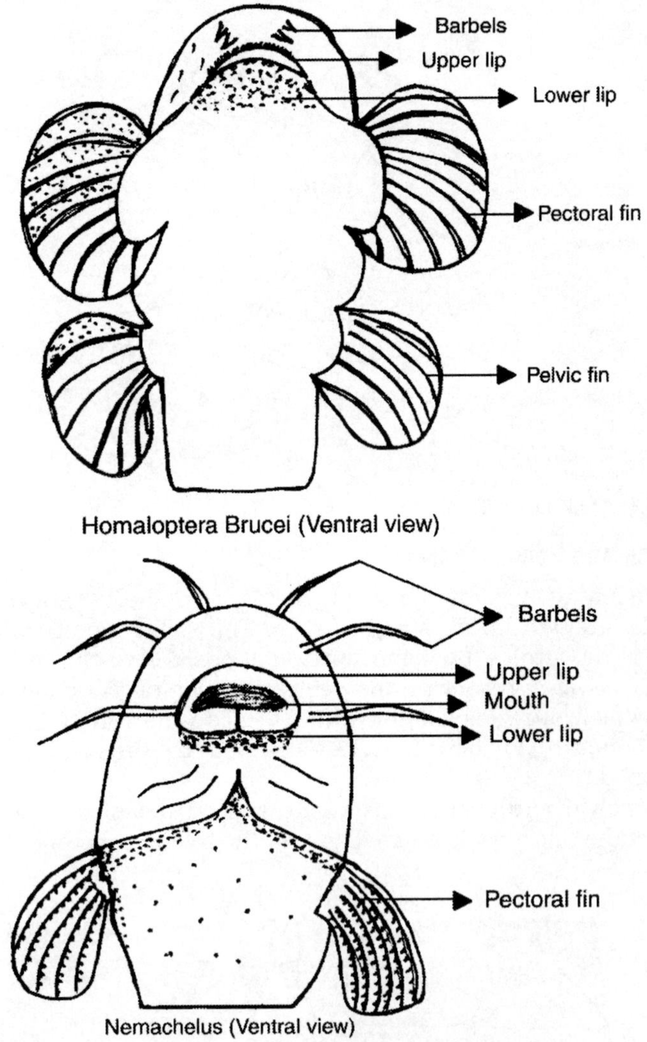

Fig. 6.3: Homaloptera brucei (ventral view) and Nemacheilus

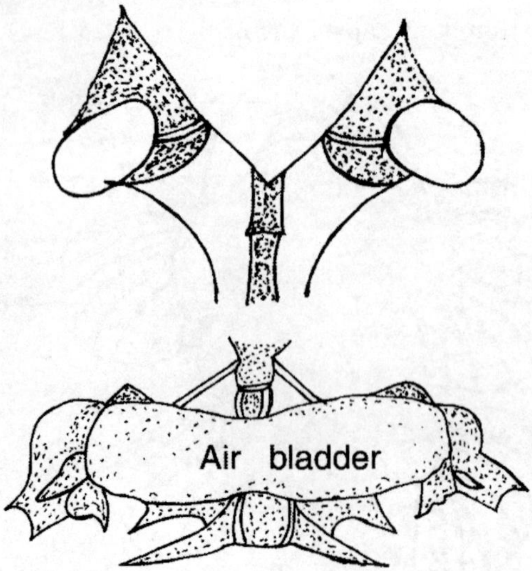

Fig. 6.4: Air bladder of Glyptothorax

Pectoral and Pelvic Girdles

The fusion of several bones takes place in the pectoral girdle to provide strength in hill stream fishes. The slight modification in the shape of the girdles, owing to horizontal position of the fins. Keel like projections are present on the ventral surface of the inner clavicular bone, which provides suitable attachment to their muscles. Adductor and abductor (Fig.6.5) muscles attached to the girdles are well developed.

In pelvic girdle, there are some special muscles which help in keeping the fins closely penned against the rocks and stones.

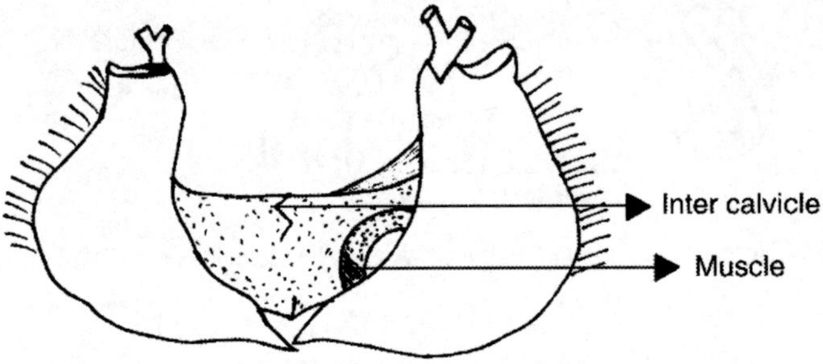

Fig. 6.5: Pectoral girdle of Glyptothorax (dorsal view)

Gill Opening and Branchiostegal Ray's

Gill opening are flat and restricted to sides only. The openings are reduced in size, with the result that probably the respiration suffers. The water is well aerated and can be retained in the branchial chamber and the pectoral fins are kept in constant motion. Thus the problem of respiratory difficulty is solved. In Garra the gill opening are wider.

In *Gyrinocheilus*, the external gill opening consist of an upper and a lower part. The water is drawn in through a kind of siphon in the upper part and allowed to flow over, the gills before being expelled out through the lower half on the gill opening.

In *Homalopteroides*, the breathing is quite irregular, at times. The water is stored in the cavities of throat and the gills chambers, for respiration, while the mouth remains free to engage in adhesion or feeding.

The restrictions of the gill opening to the sides may effect respiration of fishes. When the fish is feeding algal slime or attached to rocks and stones, the respiration may become difficult, but the following factors helps in respiration.

1. Water in the hills is better oxygenated.
2. The gill openings are small, the fish is able to retain water in the brachial chamber for a longer time.
3. The inner rays of the pectoral fins are kept in constant motion and help in respiration by forcing the water in and out of the gill opening.

The branchiostegal rays and membranes are reduced in hill stream fishes. This may be due to the backward shifting of the mouth and reduction of gill slits.

Modification of Alimentary Canal

Hill stream fishes are carnivorous or herbivorous and the alimentary canal is adapted with particular feeding habit of the fish. In carnivorous fishes, alimentary canal is short and straight tube but in herbivorous fishes, the alimentary canal is typically long or coiled.

The adaptations in both, the nature of diet, the longer intestine is found in fish. The food is poor in nutritional substances but poor developed folds as muscosa is compensated by a long intestine which sufficiently increases the absorbing surface.

Skin and other Modifications

The skin is modified in several ways to form organs of adhesion. In some, at the under surface of the paired fins, it gets thickened to form cushion like pads, grooves, and ridges which help them in clinging to rocks and stones. In others the skin instead of remaining smooth, forms grooves and ridges at the anterior and ventral side which help the animal in adhesion. In *Glyptothorax* pectinopterus there is a thoracic adhesive pad consist of an inverted U-shaped area lying posterior to the region of branchiostegal rays and between the pectoral fins. In Garra the adhesive organs are in the form of suctorial disc. In Lagusia and Erethistes striations are well marked on the chest. In Pseudo Deshone sulcatus adhesive organs consist of transfers ridges and folds.

The vary remarkable modifications takes places in the globe fish (Tetraodon) of East India. It is voracious and predatory inspite of its small size and is able to inflate its body to avoid being carried away be the rapid streams; it creeps through narrow openings between stones or behind empty shells of snails which it has filled up and swells out like a balloon. It is hen so well jammed between the stones that is remains just where it wants to without much effort.

Adhesive Organ

The skin between the pectoral fin and mouth is modified in hill steam fishes to form the adhesive organ.

In Cyprinids, the skins of the ventral side of the outer rays of the paired fins is greatly thickened so as to form cushion like structure. The cushioned pads help the fishes in securing the attachment to rocks and stones. The Asiatic cat fishes *e.g.* the *Glyptostemum* bear a serine of ridges on the under surface of the body, which act as the frictional devices in preventing the fishes from slipping in fast current of water and is supported and strengthened by cartilaginous rays lying at the bases of the paired fins.

In genus Erethistes are included a number of small fishes having smooth and flattened under surface, but in *E. elongatia* well developed striations on the chest and belly. In genus Laguvia, striations are present on the chest only and the belly is quite smooth. In *Glyptothorax* a well-defined U-shaped or V-shaped adhesive apparatus formed of the skin is present between the bases of the pectoral fins but only the epidermis is modified for adhesion.

In *Glypostemum* the chest is devoid of adhesive apparatus but the skin on the under surface of the pectoral spine is striated.

In the outer layer is present spines, the spines directed backwards the cuticle having the single row of squarish cells the produced into a beak like process all have a single nucleus and protoplasmic process are directed backwards. Below this layer are scattered cells in several rows. These cells are smaller in size. The epidermis is formed of closely packed, columnar epithelium cells. The dermis is composed of loose connective tissue and numerous blood capillaries. The cuticular spines all projecting the posterior side behave like pegs and help in securing for firm adhesion of fish. (Fig 6.6)

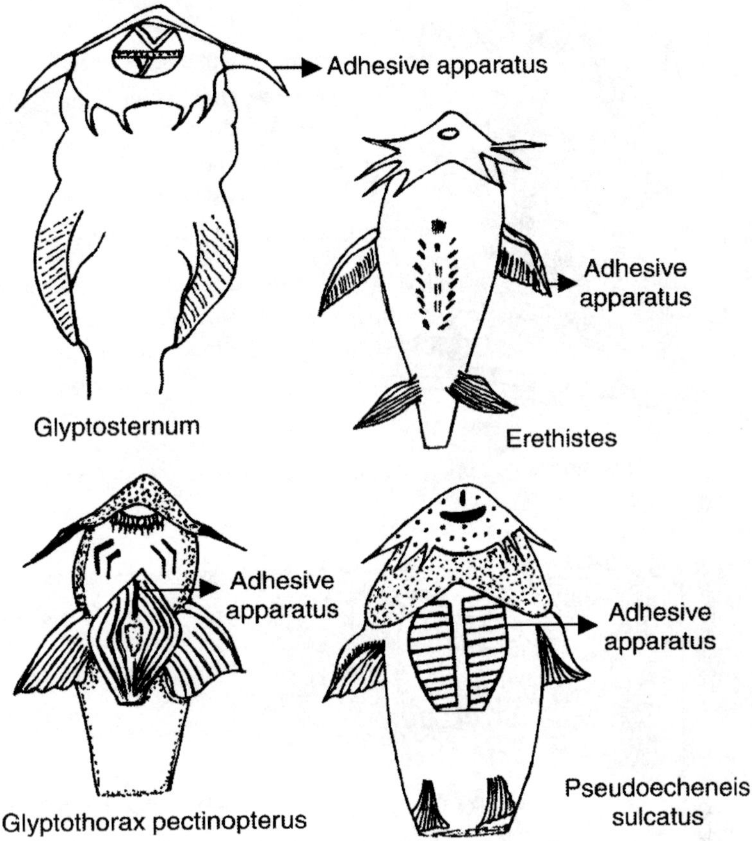

Fig. 6.6: Adhesive apparatus of Hill stream Fishes

Histology of the Adhesive Apparatus

The skin forming the adhesive apparatus is specialized in many ways. A section passing through a simple structure like thickened skin shows the following structure. The skin is divided two layers epidermis and dermis. The epidermis is formed of several tiers of cells of different shape and size and resting on loose connective tissue. The outer epidermal layer is modified into stiff and strong spine like processes. These spine rest on homogenous layer of protoplasm. They constitute about one third part of the thickness of the epidermis. Below the spines there are several layers of rectangular cells, each cell having the central nucleus. Below these cells lies a layer of columnar cells followed by loose connective tissue.

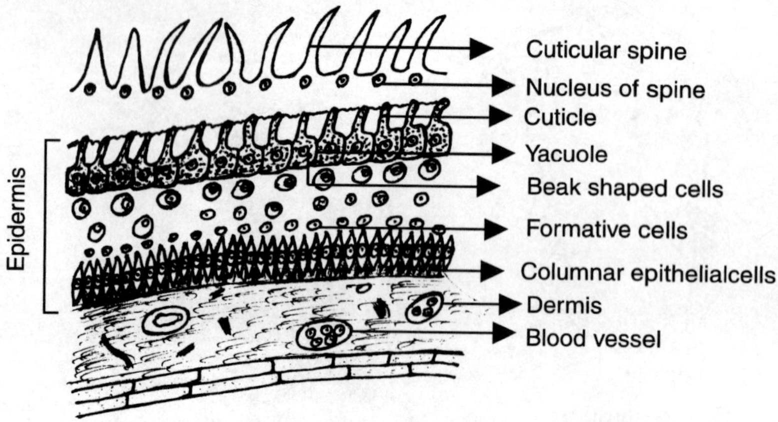

T.S. Adhesive apparatus of Glyptothorax

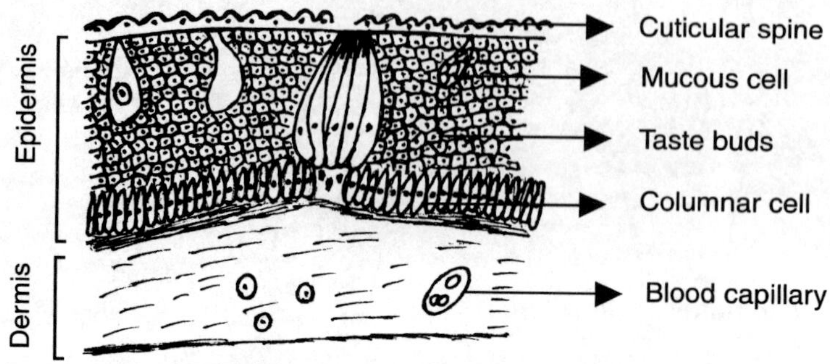

Fig. 6.7: T. S. Lip of Glyptothorax

The upper and lower lips of skin as well mandibular region forms papilla to help in adhesion. The epidermis shows modifications similar to the thoracic apparatus. The cuticle is thickened and develops into short spiny projections. The epidermal cells are oval and granular. The lower layer is composed of columnar cells and pointed ends. The epidermis to have large mucous secreting cells and taste buds. (Fig. 6.7)

Mechanism of Adhesion

The vacuum produced by ridges and grooves of the adhesive disc, brings about attachment with rocks etc. The cuticular spines, all projecting towards the posterior side behave like pegs and help the fish in securing firm adhesion. In siluroid fishes, the ridges are pulled outwards and a sort of vacuum is formed in the grooves to attach the fish to the substratum.

In cyprinid fishes, the cuticular spines of tubercles are the chief organs of adhesion. These spiny projections penetrate the organic matter covering the substratum and acting as pegs and anchors to bring adhesion.

Paired Fins

Paired fins specially the pectorals, are modified in the hill stream fishes. The outer rays of the paired fin are modified for adhesion and the number of inner rays is, therefore increased. The outer rays of the paired fins become thick and flat. The fins are shifted outwards and come to lie horizontally along the sides of the body, so as to facilitate adhesion to rocks and stones. The outer rays which are thick and flat serve for adhesion, while the inner rays are directed upwards against the sides of the body and are kept in motion. In this way the inner rays may help in respiration also. (Fig. 6.8)

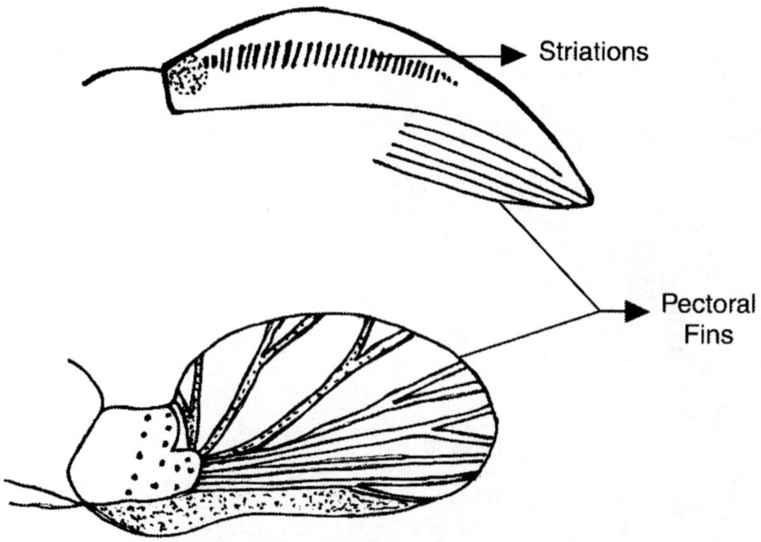

Fig. 6.8: Pectoral Fin showing reflected inner rays.

CONCLUSIONS

The food is available in hill streams in the form of algal slime but only those species can utilize it which have their jaws modified for rasping this food.

As regards their safety there is very little competition in the hill stream due to the absence of larger predatory animals.

The shelter is available below stones rocks etc. The fishes living in the sluggish water of the plains slowly ascended the hill streams in search of food and shelter.

Gradually thus, acquired characters making them specialized to survive in changed conditions.

Chapter 7

Age and Growth of Fishes

INTRODUCTION

The knowledge of the age and rate of growth in fishes is extremely useful in fishery management and both are very closely related. The age and growth are of great biological significance. Age gives the idea of life activities of an animal in relation to its size and growth. Growth is defined as an increase in any dimension of any organism in relation to time. The problem such as longevity, age of sexual maturity, age of spawning, age of catchable size, rate of growth and many others depend on the ability of fishery biologists in accurately deciphering the age of individuals. Age and growth is an important aspects in the study of natural history of fishes. There are many practical application to the study of growth rate, few of them are mentioned below.

1. The age of fishes can be determined and one can know when it reaches maturity.
2. One can know the time when a fresh stock of young fishes will begin reproduction.
3. By the comparison of the rates of growth, an assessment about the suitability of environment can be marked.
4. The estimation of the growth rate makes it possible to select suitable nets of definite mesh for catching fishes of a particular age groups.

DETERMINATION OF AGE OF FISH

Age determination is an old practice, some of the important methods of age determination are mentioned below.

Otolith Method

The otoliths are found in the internal ear of the fish and helps in balancing the body. It is called ear stone and bears annulation formed by the regular deposition of calcium salts. The annulations are formed when there is a differential growth rate in different season. The fish grows rapidly in summer in comparison to winter season. When the growth is slow the formation of annulation will be slowed and if fast the annulation will be formed fast. The annulation varies in different fishes. By observing annulations like the rings in the transverse section as a stem, age of the fish can be calculated by observing and counting the annulations. For example; the otoliths are larger structure in Gobidae family and small and thin in family Cyprinidae. For studying otolith, the fish is killed and then dissected to get the otolith.

Preparation of Otolith for Age Determination

Grinding is usually done by means of a special spherical apparatus called grinder. It can also be done successfully by placing the otolith on the tip of a finger and grinding on a glass plate using carbarandum powder.

In these cases where it has to be sectioned it is done by cutting in transverse plane and then examining along the sectional surface. Polishing is generally done by various liquids of high refractive index such as immersion oils, creozote etc. The otoliths thus prepared are examined under reflected light for age determination. The annulations in the otolith are correlated with the age of the fish. The otoliths have been successfully used for the age determination and the growth rate analysis.

Bone Method

Some bones such as operculum, vertebra, supra occipital, scapula etc also bear annulations with the growth of the fish. These structures also grow and possess deposition of the calcium salts. The growth is different in different seasons. The annual rings of these structures also help in determining the age of fish.

The center of vertebra are also helpful in denoting the age of fish. The centrum of the vertebra has layers or rings which are used in the determination of age. The tissues attached to centrum are removed by solution made up of 0.7% pepsin in 0.2% of hydrochloric acid for the study of the annual growth Rings can be observed on the face of centrum under the low power of microscope. (Fig. 7.1)

Known age Method

The estimation of growth rate by this method should be of known age *i.e.* the time of fish breeding should be known. The young fish- larvae are placed in tank or pond and left there for two or three seasons. The periodic samples are taken out and growth is measured. A growth rate curve is obtained by taking periodic samples and measuring them. One should always mark or tag and release them in their natural environment like sea, river, pond etc. Fishes are recaptured, at regular intervals and idea of rate of growth in relation to the time is gathered. This method is of limited practical value because of high cost and time consumed.

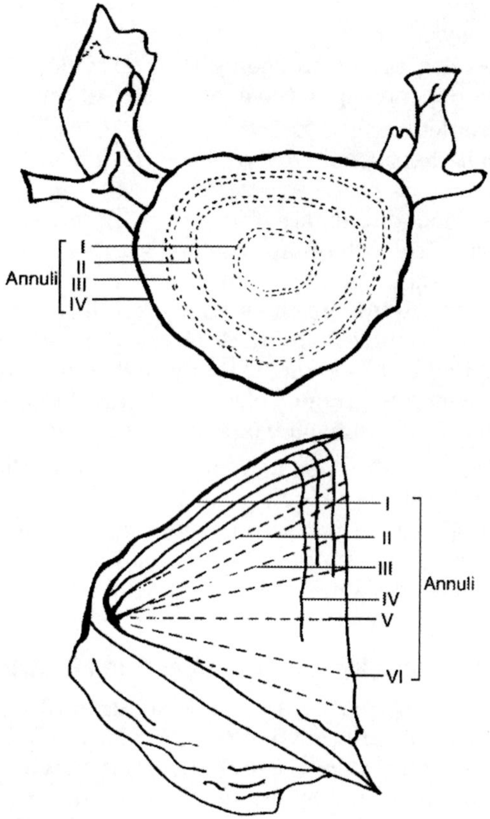

Fig. 7.1: Opercular bone

Scale Method

The scale structure and development is very useful in the inter preparation of the growth zones. The scales are collected from dorsal, ventral lateral & caudal regions. Each scale possesses a focus, around which concentric rings denotes the age of the fish. The rings are circuli, grooves, radii and anualli.

1. ***Focus*** is the central, small, clear area and displaced from the centre of the scale.
2. ***Circuli*** are concentric marks around the focus. They run parallel to each other and at more or less regular intervals.
3. ***Grooves*** alternate with ridges of circuli and are responsible for the regular spacings between them.
4. ***Radii*** are grooves disposed radially. Radii cut the circuli as they

fall in their ways.

5. ***Annuli or rings*** appear in ageing fish, as wide circular troughs, their number depends upon the age in years of the individual.

The one complete ring is called as annual ring another which is incomplete is called seasonal ring which is found in between annual rings.

The grooves and circuli are the marks of growth activity and secretion. Pattern of the osteoblasts as they secrete the material of the scale around the focus. The large number of circuli and grooves are formed in one year. Circuli are the actual sites of heavy deposition of bone material (ichthylepidin) and their height depends on the amount of calcification. Radil are devoid of body material. Annuli represent the period of slow growth in one year. Age determination according to annuli are correct if following facts are in correlation.

1. The calculated age from scale study shows correlation with the size of the individual.
2. The calculated age groups from scale study corresponds with length-frequency distribution of the individuals.
3. The assessed age from scale method is similar to the age determined by other methods.

Preparation of the Scales for the Study of Age and Growth

The scales are generally placed in a weak solution of caustic soda for some time. If the scales are thin and transparent it is not necessary to treat them with caustic soda. After the treatment with caustic soda, the scales are washed and hydrated, stained and mounted. These scales are stained with borax carmine or alizarine or alizarine sulphonate and then the growth rings will denote the age of fish. It can be studied under a microscope (Fig. 7.2). Scale method of age determination is applicable with a fair degree of certainty for many fishes such as carps, salmons, herrings, cods etc. An individual is assigned to its age and growth in the following manner which is more or less universally accepted.

Number of annuli on the scales		Estimated age	Designation of are group
1.	None.	Less than one year old	0+
2.	One	One year old.	1
3.	One	More than one year old	1+
4.	Two	Two year old.	2
5.	Two	More than two year old but less than 3 years old.	2+

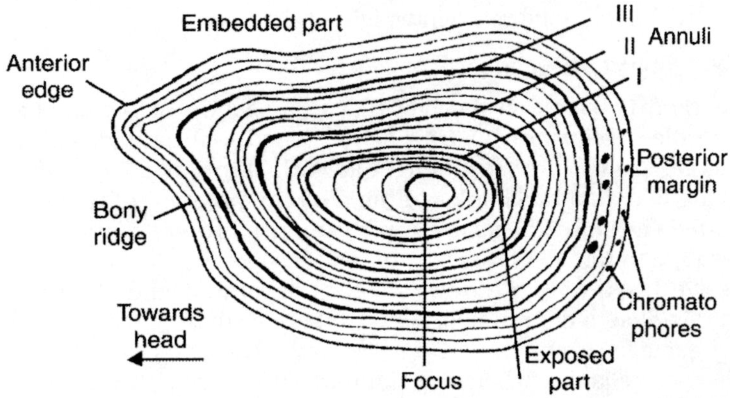

Fig. 7.2: Scale method Fish Scale showing growth rings.

Difficulties however arise when more than one "Annulus" are laid down in one year regenerated scales show fewer annuli than normal ones.

Scale Reading

In scale reading the annulus is of different type. A true annulus may be easily spotted where narrowly spaced circuli are cut across by complete circulus, at the anterolateral corners of the scale. The total number of annuli read on the scale indicate the age equal to that of years. The annuli formation and interpretation becomes difficult. It is necessary to check such readings by comparing these with normal standard growth pattern. The common abnormalities include bone.

False Annulus

It appears similar to true annulus but it takes position on the scale closer to the true annulus of the preceding year than the normal annulus for the next year will appear in case of normal growth.

Skipped Annulus

This type as annulus by position, coincides with the annulus of the preceding year. With no normal circuli forming in between this abnormal formation which is due to no growth during one growing season either in length or weight.

Overlapping Annulus resembles Skipped Annulus

In its formation it is separated from the preceding years annulus by few circuli. This abnormal formation takes place because there is slight growth during one growing season. The length of the body in this case may show some increase from the length of the preceding year but without any increase in weight of the body.

Length Frequency Method

This method is based on the fact that in a sample of individual of a given species, frequency analysis shows lengths of individuals of one age varying around the mean length according to normal distribution. In this method data of samples of the entire population of that species are plotted (normal bell-curves graph) and peaks were counted which represent clumping of fishes of successive ages according to successive lengths. Age groups are thus separated and the age of a given individual is determined. This method is suitable for application to younger age group (first 2 to 4 years). For older age groups, it becomes unreliable when overlapping of length frequencies in individuals of different age. This is because of selectivity of the method or year used hardly to ensures an is taken in to account ideal sample *i.e.* made up of representatives of all the size and age groups of the population in a restricted period of time. This method is still useful in checking the results obtained from methods of age determination by scale method. Peterson was the first to introduce this method of estimation of growth rate. This method has been reported since 19th Century. This method, however, has got certain limitations and pre-requisites upon which good results depend.

Limitation

1. This method is applicable to fishes having longer life span. For short lived fishes it is difficult to determine age, secondly size frequency is not the criteria for determining age group.
2. It is not suitable to the older age groups where the growth rate becomes slow and the number of individual are also less.
3. It is not applicable to the fishes with non-seasonal breeding cycles.
4. This method does not give any indication of the various age groups in those fishes where the individual varies much in the growth rate.

Pre-Requisites

1. The samples of fish, which are used for the analysis should be as large as possible. The larger the sample the greater is the accuracy.
2. The samples should be collected in a restricted period of time probably in a single day. According to Michael Graham (1921) this method is applicable invariably to the fishes of 3 to 4 years of age.
3. The samples should be a good representative of all sizes and age groups of a population. The method is generally used for

corroborating the other methods of age and growth estimation and is no principally applied as the only method for the estimation of age and growth rate.

Pectoral Spine or Finray Method

Spines of the pectoral fins also helps in age determination. Fine transverse sections of the fin rays are cut approximately of 3 to 5 in thickness. Sections are dried carefully and slightly heated. One section is placed in glycerin and is examined under a microscope, 1 to 2% of indistinct annuli of a fin ray are compared to 15 to 20% of annuli on the scales and thus age is calculated.

RELATIONSHIP BETWEEN THE STRUCTURES USED FOR THE AGE AND SIZE

In fish the structure of otolith, vertebra, and operculum etc. grows with the increase of size of the fish. Hence the relationship between the two can be established. If the growth is constant and uniform, it can be calculated.

$$Lx = \frac{1n}{1y} \times Ly$$

Lx = The body growth to be calculated.
Ly = Final length of the fish.
1n = Length of the annual zone or ring.
Ly = Final length of the structure (Scale or otolith).

If the growth remains uniform it is said to be isogenic and if not uniform, it is called heterogenic. If the growth is slower, it is marked –ve and if faster it is +ve. The relationship follows the law of compound interest which states.

$$L = 1k$$

L = Length of the fish to be calculated.
I = Length of the otolith or scale.
K = Constant coefficient.

This formula is usually applied for the heterogenic groups.

Thus concluded that the growth of fish follows certain measurable pattern which can be interpreted for the determination of the age and growth.

Methods of Age Reading

The annulations mentioned in above methods for different structures sometimes show false rings. These rings are incomplete and do not encircle any opaque centre. This hinders in recording the correct age of the fish. For accurate reading special methods are adopted. The structure are photographed and enlarged and then studied under a microscope. For better view, these picture are projected on screen and magnification will reveal proper structure for detailed study.

Corroborative Evidence

Graham and Vanosten emphasized before attempting to interpret the zones of otolith, scales, bones etc. The corroborative evidence alongwith proper evidence of the age of fish and the annual zones can be confirmed by following methods.

1. Agreement with Size Frequency Distribution

The size of the fishes is determined first by age reading corresponding to various lengths and the mean length of each year group is calculated that is based on the size frequency corresponds with the measurement of the stock of the year. This indirect method is providing evidence and is not fully applicable to the older age groups where but overlapping is found.

2. Seasonal Record of Zone Formation

If one opaque and one transparent zones is formed in one year, the examination of the growing periphery otolith, bone or scale indicate the period of each zone formation of one year. There are more or two zones formed in one year, this will be indicated clearly in a different way. This is a direct method which provides conclusive evidences about the age determination of fishes.

3. Observation of Fishes of the Known Age

It is a direct method of testing the accurate age. The fishes of known age are kept in an aquarium and their structures are examined periodically. These can be compared with the structures of others for the age determination.

GROWTH OF FISHES

Growth of organism means a change in length or weight or both with increasing age. Growth is generally an increase in size due to conversion of food matter into building matter of the body. In other words most fishes do not cause growth even after sexual maturity. Growth in animals has been defined as addition of structure and flesh, hence protein element. Fish growth in particular is as great importance to man as an exploiter and potential manager of fish populations for sport.

The rate of growth varies to a large extent among the various species of fishes as they are cold blooded vertebrates (Poikilotherms). It may be the same from different localities, the same individuals in different season. The rate of growth is influenced by many physical factors. The different rate of growth, of fish is dependent on population density. Higher densities show slow growth and lower densities shows faster growth. This shows the adaptive property of the species. Faster growth results in large size and protects it against the predators, while slow growth results in smaller size and associates with increased reproductive capacity and are on the mercy of predators.

Linear Growth Characteristics

During the life time of fish, growth characterizes at various periods. As far as the linear growth is concerned it is at the maximum during the period preceding the onset of maturity. As the maturity is attained, the growth starts decreasing. The food consumed is utilized for the maintenance of the growth and maturation of gonads. Each species has growth characteristics depending on following facts.

- Optimum temperature for metabolism and growth.
- Adequate and correct food in quantity and quality.
- Presence of vitamins in food (Vitamin C)
- Periodicity of season of fast and slow growth in a year. Any change in these factor influences the growth.

Determination of Linear Growth of Length

The scales, otolith, operculum, vertebra etc. present in fishes bear rings as evidences not only of the age but of growth year after year. This gives the idea of growth after every year and also of black-calculating body lengths. Annuli of the scales represent areas of retarded growth and are formed at the same time each year. Some of the methods of estimating of lengths is given below:

Body Scale Relationship Method

In this method, growth of scales and the body are at the same proportion rate. Because of this fact the fish attains some body length before the formations of scales. The formula used is expressed.

$$\frac{S}{S} = \frac{L}{L} \text{ or } L \quad \frac{S}{S} \times L$$

S^+ = Length of scale radius to annulus.

S = Length of total scale radius.

L^+ = Length of fish when annulus was formed.

L = Length of fish when scale sample was obtained.

Suppose: L = 200 mm.

S = 5 mm.

S^+ = 2 mm

Then the length of the fish at the time of first annulus was:

$$L^{+} = \frac{S}{S} \times L \quad \frac{2x200}{5} \quad 80mm.$$

Dahl-Lea-Method

This method holds that there is a curvilinear (a sigmoid curve) relationship between body length and scales length and scales length and not a straight line relationship. Therefore, a correction factor is used, and the past length is calculated as:

$$L = CS$$

L = body length.

S = Scale length.

C = a constant.

$$LogL = ogC + nlogS.$$

Where C and n-constant from data.

Several modification of Dahl-lea-method have been made in order that it can be applied to a particular species.

Length Weight Relationship in Fishes

When the accretion of fish length is plotted against time throughout the life span, the growth rate is represented by a curve.

Seasonal fluctuations in growth are often disregarded and usually empirical data on growth are available only for the post embryonic periods of the life history.

The length of a fish is often easily obtained than its weight or age. The following relation between the length and the weight of a fish.

$$W = a. L^{n}$$

a = constant.

n = fluctuates between 2.5 and 4.
Most frequently near 3.

The growth represent an increase in three dimensions whereas length measurements are taken in one dimension. This relationship also lends itself to the comparison of individuals within and between different fish populations. The value sought then is the constant d, then expressed as ponderal index or condition factor k in the formula.

$$k = W/L^{3}$$

Weight of fish is a function of its length and is expressed as:

$$w = KL^{3}$$

w = weight.

1 = length.

k = Constant.

Condition of Fish

Condition of fish is expressed by the ratio of length to weight of the fish at any given movement. The fish have a better condition, when this ratio is large. The measurements of body ratio of fish with regard to the length and height, as fish would show that the same fish is deep bodied and have a larger ratio of length to weight. When it is elongated the condition of fish is expressed by the Fulton's formula:

$$Q = \frac{W.100}{l^3}$$

Q = coefficient of conditions.
W = weight of fish.
L = fish length.

Coefficient of Condition or Condition Factor:

The coefficient of condition is an expression of the condition in which individual fish is or has been during a certain period. The value is useful and explained here.

1. Differences among individuals of the same length.
2. Differences arising from seasonal changes in relation to the age and sex of the fish.
3. Differences between conditions of the individuals of same species in different waters and so on. The conditions factor remains constant so long as length, weight proportion, as manifest by shape remains unchanged not withstanding how large the fish becomes. The alteration in this proportion is bound to alter the constant value of condition factor.
4. Rearing the fish in a controlled environment: This method is a direct method for determining the growth of fish. A fish or eggs or larvae of known age are placed in a tank or in a cage in a river. Its length and body weight is measured at intervals of time to calculate growth rate. This method is specially useful for determining growth of cultured fishes as feeding rate, temperature and other environmental factor can be kept under control.
5. Radio carbon uptake method: Fish scales taken out from the epidermis of a live fish are incubated in a medium containing the amino acid glycine made ratio radioactive by C^{14}. The rate at which the C^{14} is incorporated into the scale after incubation of less than four hours is measured by the level of beta radiations emitted by the scale. Increased incorporation of C^{14} by the scales means factor growth rate.
6. RNA DNA Ratio Method: This is the method to measure protein synthesis rate, giving the index of growth rate. DNA contents is constant per cell, but RNA content is a function of cell protein

synthesis rate. The DNA and RNA ratio shows the rate per number of cells in a tissue sample. The ratio correlates with measured weight gains in some fishes.

Growth in animals is not constant for all the years. The young grow at a faster rate and this goes on decreasing as the animal advances in age until there is no further growth. When there is no growth it is called "length at infinity" and is designated by L. But a fish do not attain this length as they die due to predation, disease or fishing.

Regulation of Growth

Growth of fish is regulated by the growth hormone which is synthesized in the pituitary gland.

Hypophysectomy results in the cessation of growth in various species, while injections of mammalian growth hormones brings about increase in growth rate. This may be due to enhancement in the food conversion rate by stimulating stored fat mobilization, protein synthesis and insulin secretion. Increase in the growth is achieved by thyroid hormone, triiodothyronine (T_3). Addition of T_3 increases growth rate and food consumption in certain species like salmon and trout. Lone and Matty (1982) observed that androgens brings about improve in food conversion efficiency of common carp and added to their diet.

Factors affecting Growth Rate

The growth rate in fishes is highly variable and depends upon number of environment factors as, temperature, amount of dissolved oxygen, ammonia, salinity, photoperiod, degree of competition, quality of food, age and state of maturity of the fish. The temperature is the most important factor but other factors such as dissolved oxygen also influence growth rate. Eg in young salmon maximum growth rate is at 15°C. The fish when deprived of 'extra' aerobic energy for growth, the reproduction falls below certain level, when ammonia present in high concentration the growth rate is slows down. Growth rate depends on the population density also. In higher densities growth rate is slow and in lower densities growth rate is increased.

CONCLUSION

By Studying age and growth of a species, various growth parameters such as increase in absolute length, rate of linear growth, average size, rate of increase in weight, growth constant etc. can be calculated. These parameters are useful while comparing growth rate of different species in the same water body, or of the same species from different water bodies. Such studies also indicate 'good and bad' years of growth in fish life.

The value of growth constant and average growth constant in a species show the existence of three distend phases of life: sexual immaturity, sexual maturity and old age. Such studies are extremely useful in the fishery management.

Chapter 8

Fish Preservation and Processing

INTRODUCTION

Fish, an excellent food for mankind is more easily perishable commodity and is spoiled if it is not properly preserved. To prevent spoilage of fish, some form of preservation is necessary. Preservation means, keeping the fish, after it has landed in a condition whole some and fit for human consumption for a short period or for a longer period of over a few months. Before preservation, fish is properly cleaned. The slime, faeces, blood stains and foreign material are washed with water. Body cavity of the large fishes is also properly washed. During the period of preservation the fish is kept as fresh as possible with minimum losses in flavour, taste, odour, form, nutritive value, weight and digestibility of flesh. This preservation should cover the entire period from the time of capture of fish to its sale at the retailer's counter including any storage time. The methods used for the preservation of flesh, includes conservation by cold and also salting, drying, smoking, canning etc. The product obtained by drying, salting and smoking is popularly known as cured fish.

The cost of preservation should, however be low to ensure profitable return. The method of preservation should be simple such as applicable on the commercial scale and suitable for climate of the place. The fish or its products, when preserved, should have good keeping quality. The poorly preserved fish could be dangerous.

Besides ensuring availability of fish in a condition fresh and fit, also helps.

1. In maintaining a steady supply of fish in the market throughout the year, including the period of poor catches.
2. In keeping a control over the hike and fall in prices, thus, stabilizing the prices even in periods of low landings or bumper catches.

Fish is an every valuable sources of protein which is easily digestible due to low percentage of connective tissue. All the essential amino acids are present in fish flesh in sufficient quantities. It is also a rich source of Iodine, Phosphorus and Vitamin B, A and D. Fish therefore, is a very valuable food.

Preservation is greatly needed, but it is relatively more difficult in tropics than in temperate places. Large geographical size, poor transport facilities and lack of funds are additional problems.

PRINCIPLES OF PRESERVATION

If the principles of preservation of fish is taken into account such measures as will keep a control on the factors which cause fish spoilage. The most important principles of preservation of fish is clean sanitation. Following are the principals involved in fish preservation.

Cleaning

The main objective is to prevent spoilage of fish from contamination and unhygienic conditions. The landed fish contains numerous microbes on the body, intestine and body cavity. There are chances of contamination with dirt, dirty water or dirty hands during handling. Poor sanitary condition and careless handling of fish in tropics increases speed of spoilage, bleeding of the fish due to injury during capture. When fish is gutted it leaves blood stains and smears leading to unhygienic conditions. Therefore, cleaning of fish greatly inhibits spoilage activity. Thus, keeping hygienic conditions at all levels of handling fishes.

Gutting

After cleaning the fishes are cut along their mid ventral side and their visceral organs are removed. By removing viscera, the bacteria in the gastro intestinal tract and the enzymes of visceral organs are removed along with it to prevent bacterial decomposition and enzymic autolysis respectively.

Raising the Temperature

Heat decomposes autolytic enzymes and thereby bring their activity to a stop. Further, moist heat under pressure kills all such bacteria and are known to cause fish spoilage. Even the spores of bacteria which are more resistant than the bacteria themselves are also destroyed. Thus, even the application of this principle is very effective in stopping all chances of spoilage and in ensuring a very long term preservation.

Lowering the Temperature

This is very effective in delaying the onset and prolongation of the duration of rigormortis and in slowing down both autolytic enzyme activity as well as bacterial multiplication. At normal room temperature prevailing in tropics (27°C to 32°C), autolysis is fairly rapid. It is well known that adenosine poly phosphates are active causing rapid break down of ATP and ADP into adenine and adenosine. Quick freezing (deep freezing) to -40°C destroys many strains of bacteria and makes the surviving ones completely inactive. Many of the latter are subject to "storage death" after some time.

Quick freezing also does not bring about deformation of flesh and at the same time controls rancidity as well.

Dehydration

Removal of water content enhances greatly the keeping quality of fish. Presence of moisture is very favourable for bacterial growth and multiplication. Loss of moisture therefore, curbs bacterial activity and even kills them. The length of keeping period of fish depends upon the degree of dehydration achieved.

Use of Salt

The effectiveness is for two reasons. Firstly salting brings about removal of moisture from fish by osmosis and secondly salt enter the tissues and increases salt concentration to saturation point. The presence of high salt concentration destroys autolytic enzymes and halts bacterial activity. Many microbes are destroyed as salt has a toxic effect upon them.

Uses of Fish Preservation

A number of drugs and chemicals are effective in preventing spoilage of fish. Some popular ones for industrial uses are mentioned below"

1. Ascorbic acid prevent rancidity in fatty fishes.
2. Chemical ice *i.e.* ice containing small concentration of sodium nitrite, may enhance keeping period of fish at this temperature by a week over the period achieved with ordinary ice.
3. Aureomycin an antibiotic is very effective in preventing fish spoilage. This antibiotic is effective even at low concentration and nearly 50% of this antibiotic is destroyed during cooking.
4. Phenolic components of smoke viz, guaiacol, cresol, pyrogallol, catechol from certain woods have antiseptic properties and are thus helpful in preventing fish spoilage. These are particularly bactericidal, resulting in heavy kill off in the bacterial load of fish.
5. Acetic acid is a powerful preservative.
6. Detergent like sodium ricinoleate in concentration of 0.2%-0.5% can destroy such bacteria as thrive in high salt concentration.

Chemical Preservation

Some of the oldest known fish preservatives as salicylic acid and boric acid used for preservation of pacific salmon. The number of other chemicals is used from time to time, though their industrial application was not preferred for one reason or the other. These include:

a) Sodium benzoate: it is good but not effective for control of bacteria at normal pH of fish flesh.

b) Carbon dioxide good but leaves behind an undesirable appearance of flesh.
c) Sodium hypochloride: good but expensive.
d) Sulphur dioxide and sodium bisulphate: good for shrimp preservation.
e) A mixture of sodium acid phosphate (3%) and sodium benzoate (0.25%): is very effective to control red discolouration of salted fish caused by some halophilic micro-organism for which nitrates and antibiotics have proved ineffective.
f) Hydrogen peroxide: good but expensive.
g) Sodium nitrite and sodium chloride: very effective for industrial application.

Nitrites of sodium or potassium are proven fish preservatives which enhance the keeping quality and have industrial application in various countries. Nitrites control *Clostridium botulinum*, *C. Sporogenes*, *Staphylococcus aureus*, *Eberthella typhosa* and similar other fish spoiling organisms.

Antibiotic Preservatives

These are rather recent; antibiotic preservatives. The antibacterial effect of tetracycline group was established by 1950, since then it is of great industrial usage in various countries. The common antibiotics used are aureomycin, terramycin and chloromycetin, among which the aureomycin is most effective and therefore popular. Aureomycin is strongly antibacterial in effect. The fish when stored at 0°C in sea water and the antibiotic used after the capture of fish gave best results. At a concentration of 10 kg/gm or ml, it improves tremendously the keeping quality of stored fish. Many countries have recommended safe concentrations in which the antibiotic should be used (5ppm in U.S.A.).

Besides the tetracycline antibiotics, some others are also useful. Rimocidin is well known for its antifungal effect against yeast and mold spoiling the stored fish. This antibiotic is however used in combination with aureomycin sorbic acid is equally known for its antifungal effect. Besides it is also used to control **"dun"** discolouration of slated fish caused by *Sporendonema epizooum.* Hydroxylamine is a good preservative but it is highly toxic to human health.

Exposure to Low Radiation of Gamma Rays

This helps in destroying completely both bacteria as well as the pseudomonas group of microbes, thereby preserving the wholesome fish.

Electrocuting by Ion Wind

This technique is recently developed in Britain. Electrically charged air particles, blowing as "**ion wind**" kill the microbes. Air ions are produced by a high voltage generator and discharged from a conductor maintained at higher electrical potential then earth. Negative ions formed are repelled at the negative potential, giving rise to the "**ion wind**" which is then attracted towards the object maintained at relatively positive potential. This technique has the advantage of the fish preserved is of high quality and flavour and is safe. The technique however, is still at experimental stage.

METHODS OF PRESERVATION

There are various methods of fish preservation, many of them are of local importance only.

Chilling with Ice

Lowering of the temperature to about 0°C is the most effective method of preventing putrefaction and extending the life of the fish. This is obtained by covering the fish with layers of ice. Large fishing vessels are provided with such facilities. However, ice alone is not effective for long preservation because melting water brings about a sort of bleaching of valuable flesh contents which are responsible for the flavour. But the ice is effective for short term preservation. Chilled fishes have to be stored properly at a constant temperature nearing the freezing point and the atmosphere should be saturated with water vapour to prevent dessication.

Freezing and Refrigeration

Freezing is more effective than chilling. Freezing is achieved either by using a mixture of ice and salt or refrigeration. The sub-zero temperature so obtained keeps the fish frozen throughout. During air freezing, water of the flesh separates and forms ice crystals, thus the quality of the fish is inferior. Frozen fishes are dipped in cold water stored at 10°F, Freezing in brine is a rapid process and produces better quality of fish which retains its natural appearance. In fast frozen fish, only small sized ice crystals are formed which do not damage the tissue. Freezing of fish can also be accomplished with the help of a current of cold air. In brine, freezing, fishes are dipped in cold liquid.

Preservation by freezing often causes loss of flavour and damages the tissue. To prevent rancidity the frozen fish is subjected to glazing with water or wrapped in a cover of moisture proof wax paper or cellophane. Glazing is, dipping the frozen fish in water so that finally a layer of hard ice surrounds the fish. This layer affords protection against exposure of fish fat to atmospheric oxygen, thus preventing rancidity. Refrigeration needs organization and initial capital for is establishment. But it is the only method of preservation that preserves the fish for long period during transportation.

Deep Freezing

It has the advantage of preventing preservation for long time. Deep freezing permits denaturation, retains freshness and flavour of fish to maximum. Deep freezing or quick freezing requires rapid lowering of temperature, say reaching-40°C in just half an hour. Because of this the zone of crystal formation of ice-31°C to 25°F (-0.5°C to –3.8°C) is passed in a very short time preventing formation of larger crystals which damage the tissues. It is done by air/gas freezing by brine or by metal freezing utilizing frigid air/gas liquid or metal. Before freezing, the fishes are washed and the heads of large fishes are removed. In deep freezing, bacterial action is inhibited. The fish retains its physical properties and nutritive value for several months. Cold storage preservation of fish is practiced only at limited places in India. In Mumbai and Kolkata also little cold storage facilities are available. There are few freezing plants at Mangalore, Kozhikode and Cochin. Fishes are kept overnight in the cold storage and then they are sent for marketing the next day.

Freeze Drying

This is a modification of deep freezing, completely eliminating all chances of denaturation. The deep frozen fish at-20°C is then dried by direct sublimation of ice to water vapour without any melting into liquid water. This is achieved by exposing the frozen fish to 140°C in a vacuum chamber of freezing chamber. The dried fish is packed or canned in air conditioned room. The product so obtained looks fresh, retains its flavour, colour and quality.

Canning

Canning is a complicated process that requires costly machinery and technical knowledge, hence, the products are costly. The canning is done in following steps:

1. Fresh fishes are cleaned to removed dirt and organism present on their body.
2. Fishes are gutted and cut into pieces of suitable sizes. Their head, tail, fins and viscera are removed. The pieces are cleaned and dipped in brine to remove blood from the tissues to give the proper degree of firmness and flavour to flesh.
3. Pieces are now immersed in hot water or exposed to steam to remove adhering materials, which could not be removed by

cleaning with cold water.

4. The cleaned pieces are filled in cans or jars which are still hot. The open containers containing pieces are now passed through an exhaust box receiving steam under pressure. Exhausting expands the food, drives out the air or gas bubbles from the containers flesh and provides an atmosphere of steam in upper vacant part of the container. Containers are then immediately sealed. Heating decomposes the autolytic enzymes in the flesh. Further the steam with a temperature range of 110°C to 120°C and pressure of 7 kg/6.45 Sq. cm. kills all the bacteria and their spores known to spoil the fish.
5. The sealed containers are again subjected to heat treatment to kill micro-organisms left in the flesh.
6. Heating affects the character and composition of the contained flesh. A long exposure to low temperature, provided through cold water or cooled air is considered helpful to prevent any changes in the texture or flavour.
7. Finally the containers are tested before their transportation to the market.

Sun drying

Sun drying is the simple method of curing fish and is extensively practiced in India. The total catch of sea fish is cured by drying in the sun. Bombay duck, Ribbon fish, Silver bellies and some other small marine fishes are subjected to sun drying. Fresh water fish *e.g.* Labeo, Catla, Puntius etc. are also dried in Bengal, Assam and Bihar but in vary small quantities as compared to marine fish. Drying means removal of moisture contents of the fish, so that the bacterial decomposition and enzymic autolysis does not occur, when moisture contents reduce upto 10% the fishes are not spoiled, provided they are stored in dry conditions. Fish drying is of two types.

Natural Drying

In natural drying the fishes are washed and dried in the sunshine. They are suspended on the open ground. The process has a number of disadvantage. It is slow and results in much loss, through putrefaction. It can be carried out only in dry well aerated climate receiving sunshine which is not too hot. Lastly only the thin fishes can be preserved by this method because the fat fishes have much flesh allowing bacterial decomposition to continue in deeper parts of their body. The additional disadvantage is that dried fishes require a long socking period to restore water and that the sundried fishes are not usually relished.

Artificial Drying

In artificial drying the killed fishes are cleaned, gutted and have their heads removed. They are cut lengthwise to remove large parts of their spinal column followed by washing and drying them mechanically.

Smoking

In this method fish is cleaned and brined. It is then exposed to cold or hot smoke treatment. In cold smoking first a temperature of 38°C is raised from a smokeless fire. After this heating, cold smoke at a temperature below 28°C is allowed to circulate past the fish. In case of hot smoking first a strong fire produces a temperature around 130°C. This is followed by smoking at a temperature of 40°C. The phenol content of smoke acts as an antiseptic. Good controls are necessary over density, temperature, humidity, speed of circulation, pattern of circulation and time of contact with fish. Smoke is produced by burning wood in the smoke house itself or is conducted to it through pipes from fire produced at a distance. This method is used in some parts of Orissa and Chennai. Sardines, Pomfrets, Jawfishes, Ribbon fishes, Mackerel, Hilsa and some other fishes are subjected to this process. For making smoke, only hard wood (conifer wood, saw dust etc) is used. The smoked fish are very popular in India. Smoking increases the flavour of the fish flesh.

Brining

This involves treatment of fish with a salt solution or brine in a container or cement tanks. This often is found useful in association with freezing or smoking. It also imparts a certain flavour to the product.

Salting

This is one of the oldest method for fish preservation Sodium chloride acts as preservative and prevents bacterial growth and also inactivate the enzyme. During salting partial dehydration of fish by osmosis also takes place. Salting can be dry, wet or brine. Salt and crushed ice are also spread on the fishes. For light and cold salting, conservation should take place in a cold room. For salting, the temperature should be around 16°C but for strong salting, normal atmospheric conditions are quite suitable. Heads and viscera of the large fishes are removed and deep longitudinal slits are made with knives on the flesh. Large sharks cut into several pieces before salting. Salting is carried out in many different ways.

In this method the fishes are kept in alternate layers with the salt. After a few days they are removed from the container and dried in the sun. Dry salted fishes are not relished much but they are less costly because their preservation is inexpensive.

Another way of dry salting is to fill salt in the abdominal cavity, after the process of guttation. The fishes are made to roll in salts over and over several times and then stacked between the layers of the salt. After salting, drying is done by the usual method in the dry and cool shady places.

Cold Salting

This method is done by spreading salt and crushed ice on the fish. About 22 to 26 lb of salt per ton kg of fish is usually recommended. After salting, the conservation is done in cold rooms having a temperature range of 2-3°C.

Salt may also have impurities in the form of magnesium and the calcium salts. These affect the taste of the fish and also decomposes the flesh proteins. Impurities of such kind must be removed or the pure salt should only be used to retain the taste of the fresh fish.

Salting in Brine

Salt concentration in brine varies from 20 to 30%. It is known that at the concentration of 25%, multiplication of micro-organisms stops. The killed fishes after being gutted, cleaned and washed are dipped in brine, containing the average of 16 to 20% or a maximum of 25% salt. After being so salted they are dried in cool shady places.

Wet salting or the "Ratnagiri" Method

In this method, the fish is splitted, gutted and cleaned as for dry salting. One kg. of salt is required for every three kg. of fish, half the quantity of salt is rubbed on the cut surface of the fish. On the first day they are stacked on the floor of the curing shed to a height of 1-1.2 meters. On the second day, half of the remaining salt is rubbed in and the fishes are stacked so that the top fish become the bottom one. On the third day the remaining salt is applied and the fished are stacked again. They are allowed to remain in this condition for another eight days and then removed. They are not dried afterwards. The finished product is very moist and yields to the light pressure of a finger. Curers from Goa, Ratnagiri, Viziadrug and other places in the Maharashtra resort to certain Chennai yards situated in South Canara district cure fishes by this method. The large fishes such as seer, black pomfrets, sharks, Indian salmon etc are generally treated by this method.

PRESERVATION WITH CHEMICALS

Some chemicals in different concentration have been tried as fish preservation. Preserving the pacific salmon with salicylic acid and boric acid is an old practice. Sodium benzoate, sodium hypo-chloride, carbon dioxide, sulphur dioxide, sodium bisulphate, mixture of sodium acid phosphate, sodium benzoate and hydrogen peroxide are also used as preservatives. Sodium nitrite and sodium chloride are extensively used for industrial applications. Effect of nitrite on bacterial growth is pH dependent. In Canada and some other countries nitrates of sodium and potassium are used as preservatives.

Sodium nitrite and sodium chloride in ice are very much effective.

Some antibiotics like Aureomycin, Tetramycin, Cycline, and Chloromycetin are used as preservatives. A fish when dipped in a solution of 50 ppm concentration of aureomycin remains fresh for several hours. Besides the above chemical preservatives, ascorbic acid, sodium ricinoleate (.02% to 5%), cresol and some other phenolic components are also used as preservatives.

Preservation by Exposure of Gamma Rays

An exposure of low radiation of gamma rays helps in destroying both bacteria and the pseudomonas group of microbes. Thus gamma rays can act as preservative.

OTHER METHODS

Pickling with Salt and Tamarind

This method is mostly popular in peninsular India. The mackerels are mostly cured in this manner and entire quantity produced is exported to Sri Lanka. After gutting and washing, salt with a small piece of Malabar tamarind is thrust into the abdomen of the fish. They are afterwards arranged in layers in a barrel with sprinkling of salt and tamarind between the layers. Heavy weights are placed on the top of the pile and the barrel is temporarily closed. At the end of 3 or 4 days the "self-brine" is run off through a small hole made in the bottom. The fishes are found to have shrunk considerably. They are pressed down and the barrel is filled with more fish from other barrels to make good shrinkage. When completely filled, the cover is put back on the barrel. The proportion of salt used is 45 kg per quintal of fish. This type of cure is conducted in west coast of Chennai.

Fish Pastes

Tamarind fish paste is prepared in Malabar from sea fishes. The fishes are cut into slices and salted when dry. The slices are treated with a pungent and spicy paste consisting of Chillies, mustard, garlic, tamarind etc. prepared in vinegar. A paste called "**hidal khunda**" is made by Nowgong fishermen from certain varieties of small fishes. These are placed in a hole dug in the ground and kept covered for nearly a month. The mass is then dried in the sun, powdered and packed in bamboo tubes. Sometimes a little ground pepper and some alkaline (as called "**khar**" in Assam) is mixed with the paste at this stage. In Assam a favourite fish paste called "**Shida sukti**" is made from dried fish. These are allowed to soak for a few hours in water and then packed closely in matkas, the interior of which has been well smeared with fish oil. Fish oil is also spread between the layers. The jar is sealed and buried in the ground for several months to allow the fish to develop the characteristic "**taste**". The Colombo method of curing fish by pickling is very common in Malabar and S. canara districts. Mackerel and sardines are gutted, washed and salted with little of fleshly covering of the fruits. *Garcinia cambogia* is trust into the abdominal cavity of the fish. Fish treated in this way are arranged in barrels made of wood. Salt and fruits are sprinkled between each layer and allowed to cure. It is exported to Sri Lanka.

CAUSES OF FISH SPOILAGE

Fish is spoiled due to three main factors *i.e.* chemical microbial and the enzymatic action.

1. The chemical action involves oxidation of fat contained within the fatty tissues of the fist. It is more pronounced in fat fishes for example, oil sardine, catla, trout and grass carp etc. which as a result become decolorized. The oil starts getting oxidized as soon as it comes in contact with the atmospheric air. The colour of the fat and its viscosity changes and the fish becomes strong in taste. Methods employed to prevent rancidity include application of antioxidants (like polyphenols or other viscous fluid) minimizing the exposure of fish to atmospheric air. The storage chamber is filled with either nitrogen or carbon dioxide. The fish is subjected to quick freezing to -20°C and only pure salts, free upon the impurities are used for salt curing of fish.
2. Microbial action involves bacterial decomposition of the fish flesh. These bacteria are found in the lower part of the gastro-intestinal tract and on the general body surface of fish body.

   The fish flesh having a pH 6-7 in living condition, provides a most suitable place for their growth and multiplication. Proteins, constituting its 70-90% are degraded by proteolytic organisms such as Pseudomonas, Proteus, Chromobacterium, Halobacterium and Micrococcus etc. The carbohydrates present in a small amount in the fish flesh are spoiled by carbohydrate fermenting organisms like the streptococcus, leuconostoc, and micrococcus etc. Fats constituting 5.3% of the flesh are digested by relatively few gram negative bacteria.

3. Enzymatic action is due to action of various enzymes found in the body tissue cells of fishes. They spoil the tissue by the process of autolysis and make the fish susceptible to bacterial attack. Proteinase for example candigest muscle proteins of the fish, catalase the gill spoilage and ATPase brings about *ATP* from muscle tissue in 6 to 8 hours. Autolysis of proteins results in formation of amino acids and other nitrogenous products, which on further decomposition produce ammonia, carbon dioxide, volatile basic compounds and foul smelling products.
4. Spoilage due to other factors

It is mentioned here that heavy losses of mackerel spoilage in America is from copepod, a plankton which forms their food. These copepods called **"red pepper"** are present in the stomach and cause spoilage within 24 hours even of iced fish.

SPECIAL PROBLEMS IN FISH PRESERVATION

Denaturation due to Freezing of Fish

Frozen fish is less valued than fresh fish. In short frozen fishes show **"denaturation"** of the flesh. Technically denaturation of flesh is described as a changed condition of flesh protein in which the protein cannot be extracted any longer by the known methods of solubility of various salt solutions as applied to normal fish muscle for protein extraction. The deterioration in quality of fish flesh is directly associated with the degree of insolubility, the phenomenon of frozen fish. According to Zarot-Schenzeff, (1930), freezing causes formation of crystals in the muscles. The more accepted theory is of Finn (1932) which describes that freezing causes changes in protein water salt balance in fish muscles. In normal fish muscle the combination of strength of inorganic acid and organic salts absorbed in protein is equal to the strength of 1.4% NaCl. Cooling below freezing point causes water to leave the tissue and freezes out as ice crystals. The dehydrated muscle tissue now has greater concentration of salt inside in the liquid state. The range between -20°C to -6°C is the zone of maximum ice crystal formation. For longer the time, the tissue takes to pass out of this zone during freezing is greater in the crystallization, therefore, during greater dehydration of muscle maximum damages is done at this temperature range. This involves the following.

1. Protein is more and more aggregated due to salting out of sols and gels from protein constituents particularly the less organized like actomyosin.
2. Loss of moisture to the protein micelles and gels leads to considerable dehydration resulting in change in the **"look"** of the flesh.

3. Presence of hydroscopic salt in flesh like trimethyl aminoxide and lactic acid aggravates the dehydration effect in the flesh.
4. Splitting of actomyosin into actin and myosin prolonged cold storage at higher temperatures therefore accelerate denaturation. At low temperature conditions of increased viscosity and reduced molecular diffusion develop, which do not favour protein aggregation.

Denaturation is controlled and prevented by the following methods used for industrial application.

a) Quick freezing the fish.
b) Covering the frozen fish during storage with protective wrappers like moisture proper foils, films and other like things.
c) Packing the product in sealed cans.

Problems of Industrial Processes in Fish Preservation Industries

Certain process, if carried out carelessly in unhygienic conditions or without adequate precautions do more harm than good in fish preservation. Washing is good in reducing bacterial flora of the fish surface.

Gutting of fish is aimed at preventing autolysis of flesh by the enzymes of the gut. However, gutting often helps in contamination of flesh with bacteria of the intestinal fluid. Handling of fish at different-stages such as board, on shore and in the market has the risk of helping the bacterial spoilage by contamination with unhygienic surface. Filleting of flesh particularly in fat fishes may only increase chances of rancidity. Fish autolyzed are deliberately obtained from controlled spoilage of fish under experimental conditions. Fish autolyzed are useful in studying the growth of spoilage causing bacteria as these serves as an excellent substrates.

Fish spoilage is a particular controlled spoilage preparation obtained from the activity of the bacteria Lactobacillus plantarum which causes formation of large quantities of lactic acid.

Demerits of Fish Preservation

Although the preservation and processing constitute a very important aspect of the fish industry, it has certain drawbacks as well, particularly with respect to retaining quality of fish flesh are given below:

1. Incomplete or poor preservation leads to decarboxylation of histidine into histamine. The latter and some other related substances, collectively called saurine are the common causes of food poisoning.
2. If proper hygienic measures are not taken during the processes

like washing, guttation and evisceration etc. more harm would be done to the preserved material owning to increase in the bacterial population.

3. Chilling brings about denaturation of the flesh, this is because of the ice crystals formed during chilling causing mechanical damage to the muscles, cell walls burst, structure gets deformed and the flesh loses much of flavour and taste. The flesh also becomes dehydrated and loses texture.
4. Drying reduces weight, nutritive value and the digestibility of the fish.
5. Smoking also accelerates rancidity of fat and so reduces digestibility of fat products.
6. Excess salting allows growth of salt tolerant bacteria causing "pinkeye" spoilage of fish flesh.
7. Canning leads to much loss of vitamin B, pantothenic acid, pteroylglutamic acid and vitamin 'C'.
8. Salting combined with smoking results in loss of protein about 1% to 5% due to salting and 8% to 30% due to smoking.
9. Use of antibiotics, if not carried out with caution may lead to problems of sensitization of people handling it.

PROCESSING

Fish processing includes all the processes enumerated above as cleaning, freezing, salting, drying, canning etc. Fish may also be processed into edible meals and oils that are obtained as by-products of the fish industry. Fish meal is prepared from discarded parts of the fish as gut, gills, fins etc. By processing species of small size or such species that are not popular are also converted into fish meal. Major part of the catch is consumed as fresh preserved or salted fish, but there is good scope for several byproducts of the fish industry.

Chapter 9

Transgenic Fishes

INTRODUCTION

Isolation and construction of genes responsible for desirable traits and their transfer into the germ line of broad stock may produce a quantum leap over traditional breeding and selection methods for the production of fast growing fish. Many growth hormone (GH) cDNA and full length genes of mammalian, avian and piscine origin have been transferred into fish eggs to accelerate growth. However, expression of the transgene depends mainly on the right choice of regulatory elements. The expression of these heterologous genes was found unsatisfactory for one or other of the following reasons.

(a) Undesired tissue specific expression (b) side effects leading to impairment of other functions. Hence an attempt was made to construct a transformation vector with a unique promoter Zp (Zona pellucida gene) of piscine origin to drive maximum expression of the fish growth promoting genes, yellow porgy growth hormone (Yp GHc DNA) and rainbow trout growth hormone (rt GHcDNA).

WHAT IS TRANSGENIC FISH

A transgenic fish is one which carries one or more foreign genes. The foreign genes are selectively incorporated by microinjection into the egg with a view to produce transgenic fish lines carrying such foreign genes.

The progress was made in genetic engineering to isolate eukaryotic genes in 1970s and by 1980s. It was possible to microinject such genes into animal eggs to produce transgenic animals. The technique was applied to fish much later. Some significant progress was made which had potential for application to fisheries. More than dozen transgenic fish were produced by 1989. Some examples include:-

1. Transgenic Atlantic Salmon to which antifreeze protein gene of polar flounders was transferred. This was done to promote resistance to low temperature of ice waters in the salmon so that the habitat of salmon may be extended to polar water.
2. Transgenic food fishes trout, tilapia, cat fishes etc. to which growth hormone gene of human, rat or other mammals was transferred with a view to enlarging production of such fishes.

Fish transgenics is comparatively difficult because of the tough egg chorion which impeded microinjection. A prior puncture or use of micropile (An opening in the egg surface for sperm entry during fertilization) or, local trypsin digestion has to be made for microinjection. If gene transfer by microinjection may be accomplished, integrations have been rarely achieved without which transgenics is not complete.

TRANSGENIC TECHNOLOGY

Transgenic technology involves an appropriate method of gene transfer (Fletcher and Davies 1991). Popular methods hitherto used for gene transfer in fish are:

1. Micro injection (Ozato *et al.*, 1906, Zhang et al 1990).
2. Incubation of sperm. (Khoo *et al.*, 1992).
3. Electroporation of eggs (Powers *et al.*, 1992)
4. Electroporation of sperm (Walker *et al.*, 1995)

Accordingly to Marian *et al.*, 1997, electroporation is more promising method for gene transfer. It is possible to electroporate unfertilized eggs of just fertilized eggs, but the latter is shown to be ideal, as it result in higher frequency of transgenics and low percentage of deformities. Dunham and Smither Mas (1987) describe electroporation is an alternate method in fishes like the silurids in which sperm cannot be stripped. A second objective of the present study is to subject just fertilized eggs of Zera fish to electroporation with one of the two recombinant transformation vectors ZpBpGH and ZpBrtGH.

GENES AND TRANSGENIC FISH

1. Zhu was the first to inject human insulin gene into crucian carp and loach eggs. Human growth hormone gene was the first gene that was successfully used to produce such transgenic animals.
2. The growth hormone genes of rainbow trout and salmon were sequenced and became available after 1986. The success with mammalian hormone is only 50% expression. Therefore there is an urgent need for constructing transgenes of fish origin.
3. Bacterial gene transfer into fish has become a powerful tool to study gene expression in vivo and in vitro CAT B-gal (chloramphenicol acetyltransferase, B, galactosidase) (Table No. 9.1).
4. The fish transgenic technology is applied to increase growth rate of food fish, freeze resistance genetic regulation of the development process using Zebra fish and Medaka.
5. In Asia, disease resistant and drought resistant transgenic fish production is the prime need.

Table 9.1: Advantages and limitations of using fish egg as a model for manipulation

Advantages	**Disadvantages**
Fertilization external, artificial, stripping of eggs and milt possible, maturation of gametes artificially inducible. Egg are large, numerous and easily maintained after fertilization; provide maximum scope for expression of variation. Faster embryonic development in pronucleus.	In many species, Zygote nucleus of eggs are very small and not visible hence, nucleoplasmic microinjection into oocyte nucleus of oviparous and viviparous is possible only in some species. The oocyte nucleus is over 1000 times larger than the mammals.
Fish eggs are amenable for ploidy induction, Hence possible to generate hundreds of clones from a single parent. This effectively saves one generation in the selection of homozygous mutants.	Egg coat (Chorion) is very tough and resistant.
Relatively higher totipotency of fish cells render the production of pure transgenic line a much easier task than in mammals.	

GENE MANIPULATION

For gene manipulation, physical and manual techniques are employed. Such technique requires, microneedles with less than 0.1μ tip, micropipette and holding pipette, that too the instrument is so equipped for correct micro incisión and microinjection with proper slide adjustment. Micro injection of Bovine and rat growth hormone gene in Tilapia egg has been successfully carried out. (Fig. 9.1) Tilapia egg is highly opaque and the chorion is very hard, hence it cannot be pierced through even with very fine micro needle. Therefore, foreign DNA is injected into the germinal disc through the micropyle before the starting of cell division with a micro manipulator. Microinjection of growth hormone gene in Zebra fish egg has been done directly by injecting into the cytoplasm before the first cleavage. (Fig. 9.2) Chinese have also taken trials in nuclear transplantation in crucian carp. Considerable progress has been made in the field of cloning and processing of recombinant DNA in virus, bacteria, algae, insects, sea urchins, amphibians, mammals etc. In fishes only preliminary attempts are reported in case of gold fish, medaka, loach, tilapia, trout, salmon, Zebra fish etc. However, gene manipulation, will take upto new horizon of aquaculture in passage of time and research in gene investigations.

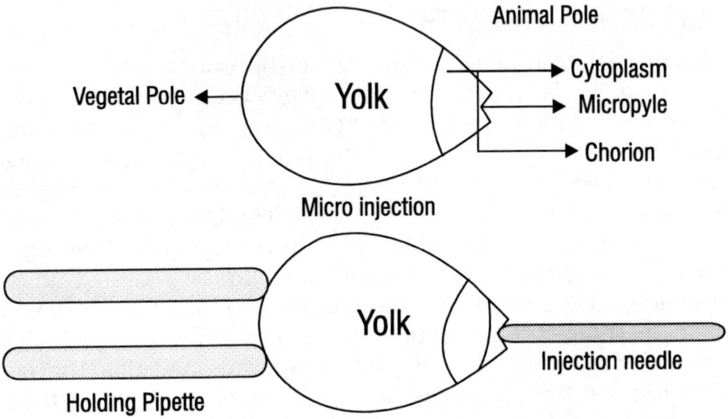

Fig. 9.1: Micromanipulation in tilapia egg.

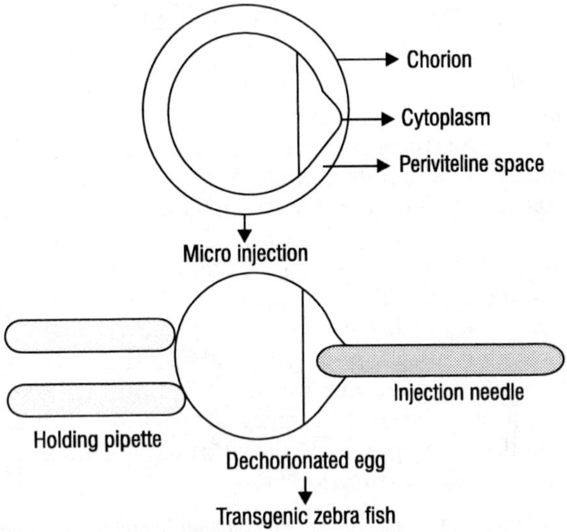

Fig. 9.2: Gene manipulation in zebra fish.

Types of Microinjection for Fish

a) Human or rat gene for growth hormone.
b) Chicken gene for delta-crystalline protein.
c) E-coli gene for B galactosidase.
d) E-coli gene for neomycin resistance.
e) Winter flounder gene for anti-freeze protein.
f) Rainbow trout gene for growth hormone.

METHOD OF GENE TRANSFER

In fishes common methods are electroporation, sperm mediated transfer, high velocity micro projectile bombardment and microinjection which are the most-preferred method. Other methods adopted are drilling of chorion, manual removal enzymic digestion with trypsin injecting through micropyle. Some of the problems faced are digestion of the injected DNA by nuclear DNA and the rotation of the egg if perivitelline space is broad. In transgenic fishes mosaicism can be done by either the cytoplasmic introduction of the transgene or by the microinjection of egg even after the pre blastoderm stage and in another method the inherent totipotency of fish eggs. The time available for microinjection is first 25 minutes and that too between fertilization and first cleavage. In forty minutes adequate cytoplasm is formed only after 15 minutes of fertilization.

The quantity of injected gene varies from human GH (Growth hormone) from 2.9 to 9.4 Kb. The largest gene introduced is chicken crystalline 14.5 kb. The volume of DNA solution is 0.2 to 20 ml. –10-15 µg/ml, which is toxic if more than 15 µg/ml is used. Number of the copies of the transgenes to the Minium is 50,000. About 1-2 million copies are contained in 10-20 ml. with 10-15 µg/ml.

There are certain reasons for sensitivity of carps and cat fishes for micro injection which includes (1) Removal of chorion (2) diameter of the needle used (3) The quantity of injected DNA (4) The quality of eggs (5) Time required for micro injection. Procedures are not fully time consuming but laborious too, at the same time it is specific and technically demanding. The production of transgenic fish by this procedure is slow and serves a bottleneck to the explanation. Explanation and realization of the full potential of this powerful technology by this method the electroporation results in 760% efficiency, although the rate of survival is low. About one million copies are required for sperm mediated transfer. Many million sperms are required to ensure fertilization.

Table 9.2: Advantages and limitations of using different methods for gene transfer in fish transgenes

Advantages	**Disadvantages**
Microinjection	
Permits precise injection at the desired location of the target egg *i.e.* nucleus. Cytoplasm permits quantitative estimation of the injected gene.	Time consuming techniques, requires skill, Limited embryonic time restricts injection to fewer eggs which introduce variations due to needle injury.
Electroporation	
Permits simultaneous entry of foreign DNA	Difficult to assess the actual quantity of

Advantages	Disadvantages
and brings about man production of fish transgenesis.	foreign gene entering the cells.
Does not requires technical expertise.	Sustains relatively higher mortality.
A more suitable method in some fish species, whose eggs are too small for microinjection.	
Sperm mediated transfer	
Absence of acrosome in fish sperm affords a greater scope for sperm mediated gene transfer.	Its usefulness is claimed in birds but disputed in fish and mammals.
High velocity micro projectile	
Bombardment permits simultaneous entry of foreign DNA and thus brings about man production of fish transgenesis.	Difficult to assess the actual quantity of foreign gene entering the cells.
	Not used much because little is known about the mechanics of the process.

Integration

As the injected DNA undergoes amplification there is degradation by DNA ase. Some extra chromosomal persistence of the transgene lead to mosaicism which is due to unequal distribution among daughter cells. There are large difference in the persisting copy number of the injected gene in tissue of transgenic trout. This is mainly due to delivery of the DNA later to two cell stage. But the integration frequency in <20% with fish genes and 740 with mammalian genes. However, the expression of the fish transgene is higher as compared to mammalian genes and 740 with mammalian genes. However, the expression of the fish transgene is higher as compared to mammalian gene. But the differences in the integration rate may be due to the tailoring of the selected gene, for hGH (human Growth Hormone) it ranged between 2.9 kb to 9.4 kb. The integration frequency was found to be 75% with 6.3% kb. and 20% with 2.9 kb.

Transmission

Due to the mosaic nature of integration, the transgenic characters are not transmitted according to the Mendelian ratios. It was found that out of 20 presumptive transformants crossed with normal fish, only one female was transgenic mosaic. The foreign DNA contained in her germ cells was 20% at about 100 copies per cell. The copy number of hGH not only varied from tissue to tissue but also from individual to individual. In Zebra fish, F_1 offspring from positive founder, parents were crossbreed to untreated partners and subsequent out crossing of their progeny F_1, progenies of the 4 founder parents displayed no transgenic character but some of their F_2 offspring proved to be transgenic.

Expression

These are two types of transgenes which are cloned (1) full length nuclear DNA and (2) complimentary DNA. Nuclear gene consist of coding sequence, promoter/enhancer sequence and natural spaces. There are three types of promoters/enhancers (1) enhancers from normal cellular genes active in most tissues (metallothionein-1: B-actin enhancer-promoter complexes) (2) enhancers active only in specific tissues (3) Viral enhancers active in many tissues, especially those normally injected by the virus. These marker genes (Luciferase), presence of mRNA by northern blot or the protein by immunoblotting.

Generalization on Expression

The viral promoters ensure more definitive expression than others like metallothionein RtGH sequence comprising Rous sarcoma virus promoter ensured expression in carp, medaka and seabream. However MT or RSV transgenic fish are not fit for human consumption. The high level of expression is observed when all fish Chimeric GH gene construct is used. The presence of adequate number of copies of the injected gene in the desired tissue organs, determines the level of expression. The heterologous genes expression is unsatisfactory due to (a) undesired tissue specific expression (b) side effects leading to impairment of other functions.

APPLICATION

Use of transgenics in selecting superior desired traits, also helped in production of super fish or super males and females. These forms owe their giant size to the introduction and incorporation of such anabolic growth promoting genes as bovine growth hormone gene, human growth hormone gene. China has produced giant loaches similar to giant mice and giant sheep produced by micro injection of human growth hormone gene.

This aspect has great fishery application from fish culture stand point. Somatic growth of farm fishes may be prolonged and made faster by introduction of bovine or human growth hormone gene.

The goal of transgenics is to produce an “ideal cultivated fish form” which may be envisaged as one having the desirable and superior traits. Such a dream fish “may be thought to combine the following traits with taste and food value.

1. Fast growth.

2. Fatty make up.
3. Greater longevity.
4. Relatively omnivorous feeding habit.
5. Higher fecundity.
6. Greater adaptability at developing stages.
7. More hardy.
8. More resistant to diseases, biocides, and pollutants.
9. Lack of fish bones and other undesirable features.

Chapter 10

Predatory and Weed Fishes

INTRODUCTION

There are numerous predators which prey on smaller fishes than on larger fishes, predators are either vertebrates or invertebrates. However among vertebrates most important groups are cat fishes belonging to siluriformes. The predatory and weed fishes particularly present in the nursery ponds attack hatchlings and prey on them. The culture of fishes during the life history is liable for predation at every stage. Beside predators in unmanaged ponds some varieties of uneconomic small fishes that are accidentally introduced with riverine carps spawn are termed weed fishes. Both these weed fishes and predatory fishes compete for food and space among the cultivable varieties of carps. Some fishes directly feed on carp hatchlings. Certain predatory fished breed, prior to breeding of major carps in pond water and feed on available planktons, with fast increase in growth and size when carp spawns are introduced, the predators are large enough to feed on them.

Weed fishes have good fecundity and attain sexual maturity in summer and breed even without rain prior to the monsoon and therefore their young ones are abundant in number during monsoon.

Table 10.1: Some examples of predatory fishes:

Clupisoma garua.	*M. cavasius.*
Silonia silondia.	*Notopterus chitala.*
Rita rita	*N. notopterus.*
Nandus nandus.	*Mastocembelus armatus.*
Bagarius bagarius.	*Pseudociana coitor.*
Pangassius pangassius.	*Channa striatus.*
Pama pama.	*C. marulius.*
Glossogobius giuris.	*C. stewarti.*
Wallago attu	*C. punctata.*
Clarias batrachus.	*C. gachua.*
Heteropneustes fossilis.	*Anabas testudineus.*
Mystus seenghala.	*Ailia coila.*
M. aor.	*A. berg.*

Fig. 10.1: *Rita rita*

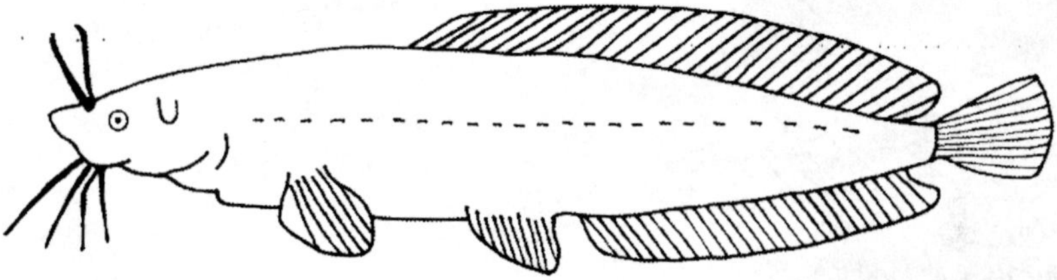

Fig. 10.2: *Clarias batrachus*

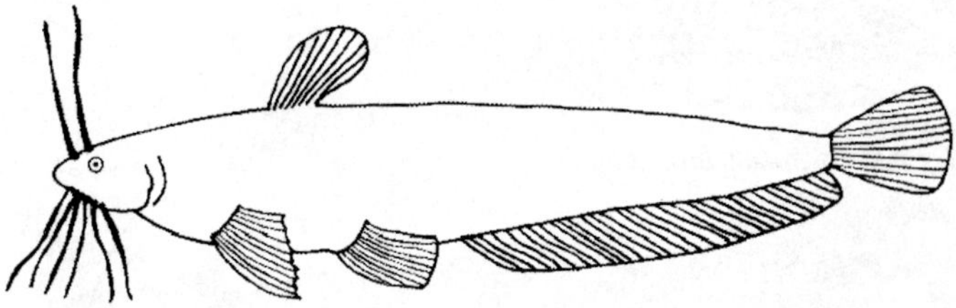

Fig. 10.3: *Heteropneustes fossilis*

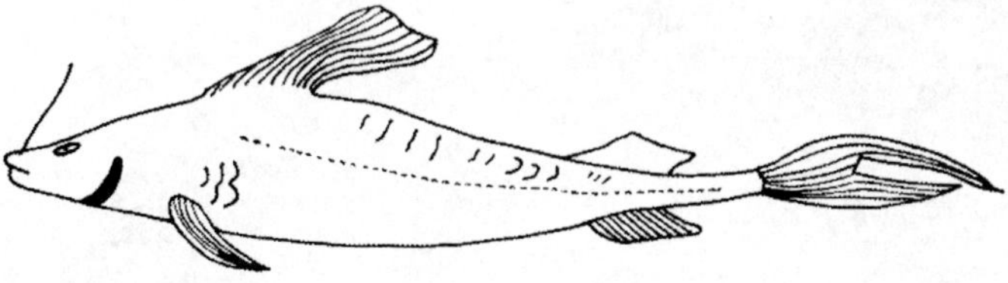

Fig. 10.4: *Mystus seenghala*

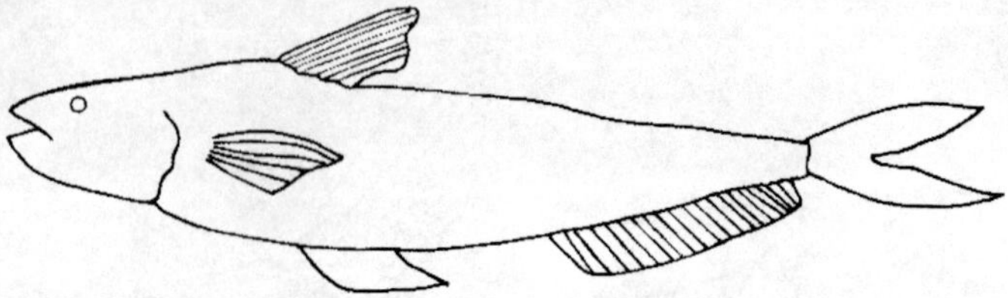

Fig. 10.5: *Pangasius pangasius*

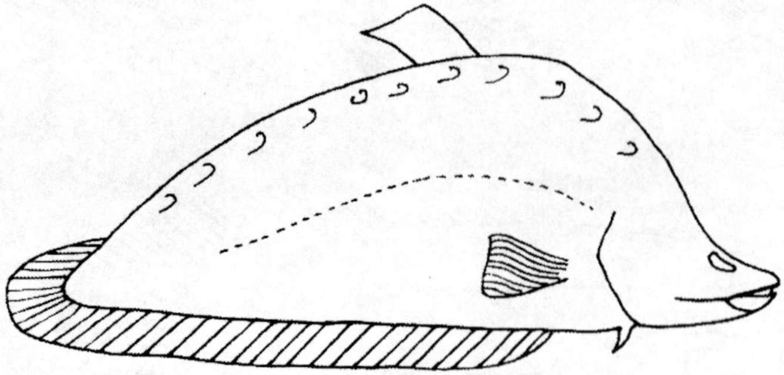

Fig. 10.6: *Notopterus chitala*

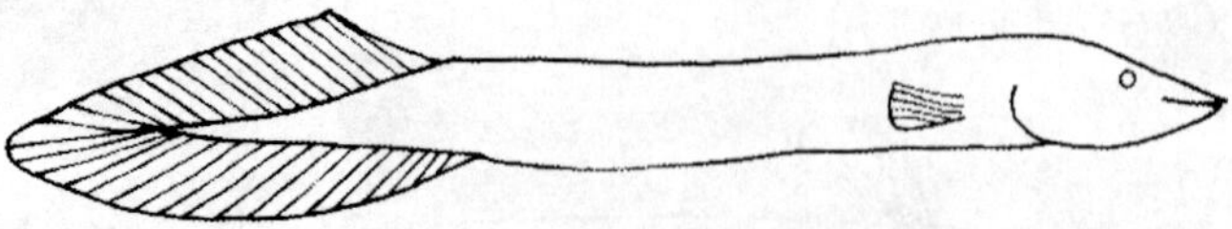

Fig. 10.7: *Mastacembelus armatus*

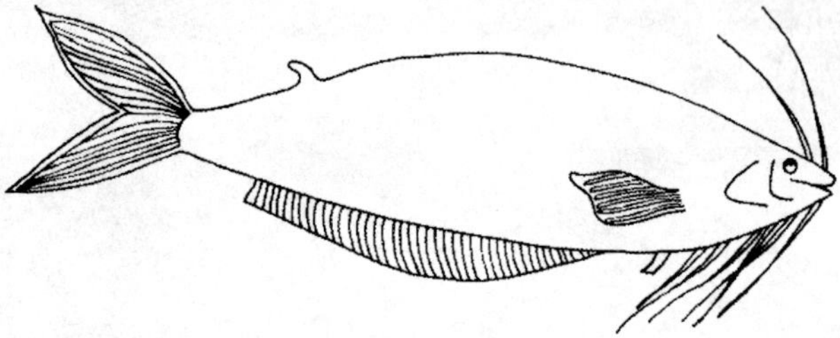

Fig. 10.8: *Ailia coila*

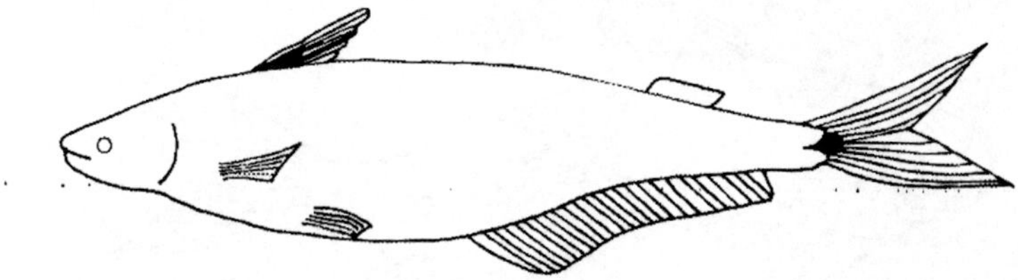

Fig. 10.9: *Silonia silondia*

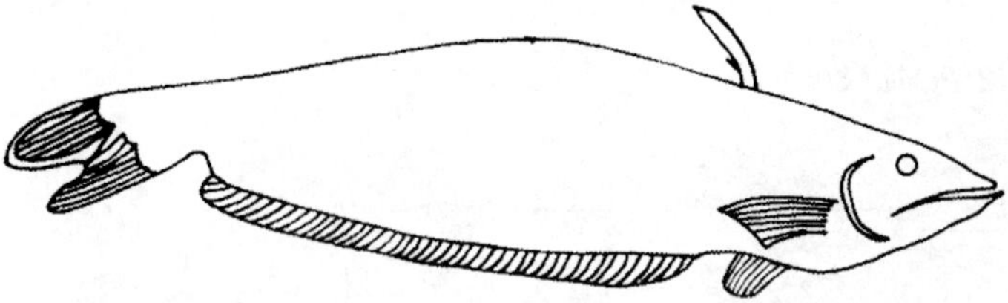

Fig. 10.10: *Wallago attu*

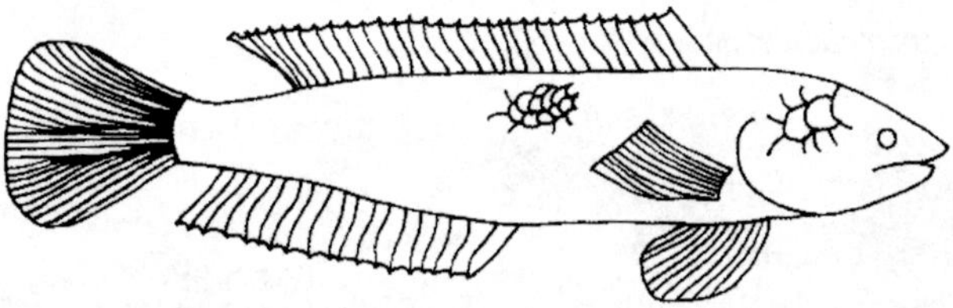

Fig. 10.11: *Channa punctatus*

Table 10.2: Some examples of weed fishes.

Ambassis ranga.	*Neomochellus zonatus.*
Aplocheilus lineatum.	*Osteobrama cotio.*
A. panchax.	*Oxygaster bacaila*
Barilius barila.	*O. phulo.*
B. bola.	*Setipina phasa.*
B. vagra.	*Punius ticto.*
Chela cochins.	*P. conchonius.*
C. laubuca.	*P. sophore.*
Esomus danricus.	*Rasbora daniconius.*
Gadusia chapra.	*Xenentodon concila*
Gonisosla manmita.	

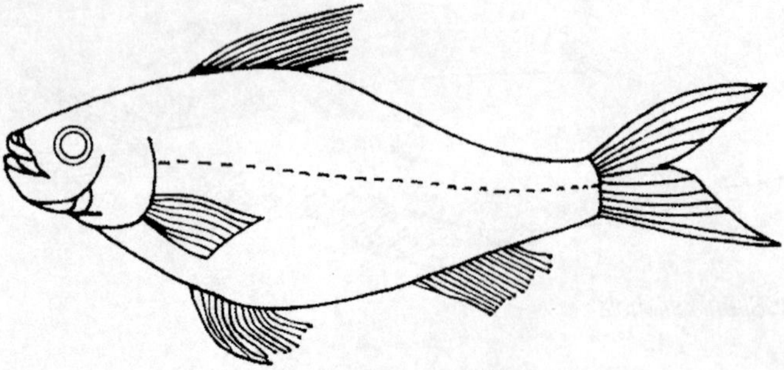

Fig. 10.12: *Puntius sophore*

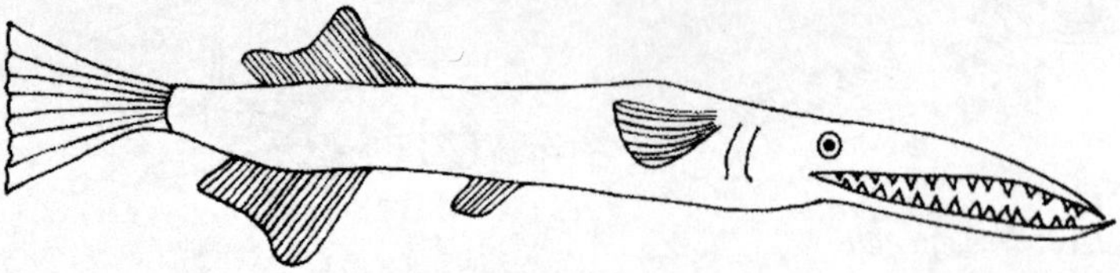

Fig. 10.13: *Xenentodon cancila*

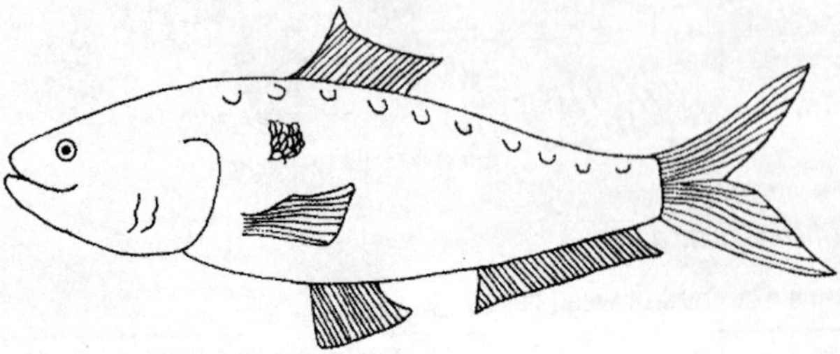

Fig. 10.14: *Gudusia chapra*

Fig. 10.15: *Oxygaster bacaila*

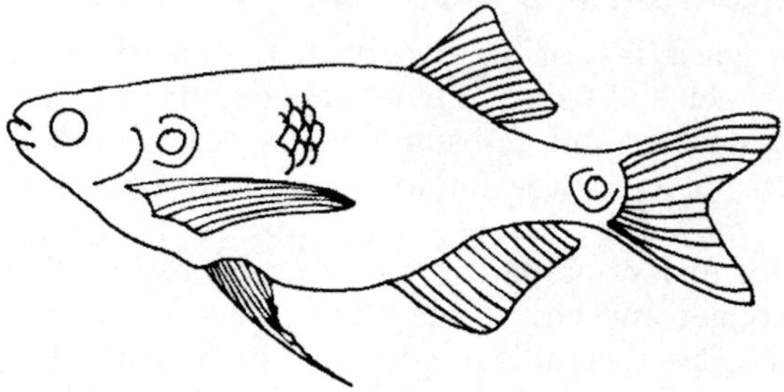

Fig. 10.16: *Chela laubuca*

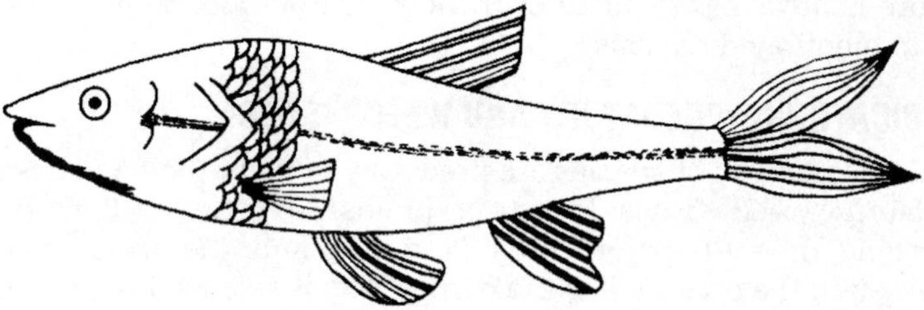

Fig. 10.17: *Rasbora daniconius*

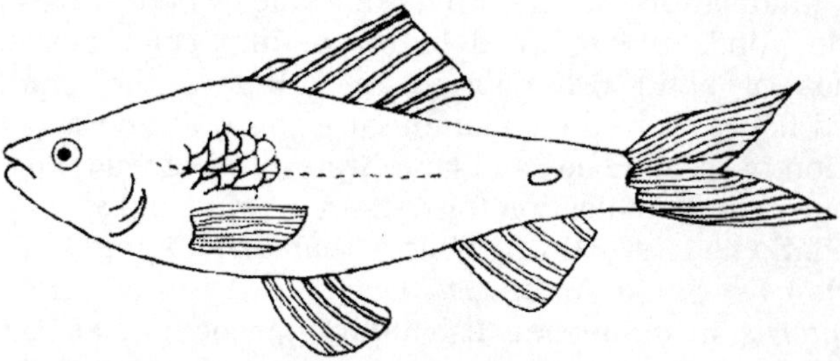

Fig. 10.18: *Puntius ticto*

Fig. 10.19: *Osteobrama cortio*

ROLE OF PREDATORY AND WEED FISHES IN PONDS

Piscivorous varieties of predatory fishes feed on minnows and culturable, species of fishes. Minnows being prolific breeders feeds on planktonic masses and consume all natural food of fishes, but minnows do not complete for food but compete for space with the culturable species of fishes. This reason necessitates complete removal or control of predatory and weed fishes in carp culture tanks.

The stronger and larger sized fishes feed on weaker and smaller fishes. Hence, it is desirable to completely eradicate the predatory and non-predatory minnows from culturable tanks. The minnows and predatory fishes are very less than the carp varieties. Therefore, carp culture is always preferable over the predators and weed fish culture due to mentioned reasons.

ERADICATON OF PREDATORY AND WEED FISHES

The large number of species are predatory and eat the fry. These some predatory weed fishes breed in ponds, a little earlier than the spawning time of major carps in rivers, and the fry feed on the plankton in the pond. When the carp spawn is released the young ones of predatory fishes become large to feed on them. The common predatory and weed fished are removed from the nurseries.

The eradication of the predatory and weed fishes from the culturable ponds is suggested to be an important measure in the techniques of pond fish culture. In old days the eradication of unwanted fishes were done mainly through hook line and with baits. Eradication of these fishes and large water bodies was uneconomical. Thus, control and eradication for nursery rearing and stocking pond is suggested. Some Ichthyologist believe that a small population of small mouth cleft fishes like *Notopterus chitala*, *Eutropiichthys vacha* and *N. notopterus* in composite fish culture is beneficial. Because the predatory fishes feed on minnows and on fry of weed fishes. Thus controlling the population of weed fishes. This view support large water bodies, control the big sized predatory fishes like *Wallago attu* or *Ophiocephalus*. This results in favourable condition for cultivable species of fishes.

METHODS OF ERADICATION

Dewatering and Desilting

Dewatering of the pond followed by the drying of soil is a successful method in eradication of unwanted fishes. But it is economical in small ponds. In heavily silted ponds desilting after dewatering is necessary. Some bottom living species like the Murrels, Singhi, Mugur,

Cuchia etc. burrow in mud. Hence, desilting helps in removing these fishes. Usually dewatering is done in summers when the depth of water in pond goes down to minimum. This method is very convenient and economically applied in the seasonal nursery and rearing ponds.

Hooks and Lines

Hooks with lines are used for catching the piscivorus fishes like Wallago attu, murrels etc. but lines and rods are used with live baits to catch such fishes. Earthworms and other small fishes are used as baits in some cases.

Netting Operations

It is a primitive method, Netting in a small water area with the help of a line meshed dragnet is effectively used for removal of the predatory fishes. For minnows more effective netting sinkers are used with the bottom rope. For the removal of predatory fishes from pond, and subsurface soil, it is suggested to stir the bottom soil by means of bamboo poles manually or by some other means. Dragnet should be operated for the removal of more number of predatory and weed fishes. Certain fishes like cuchia, murrels, singhi are removed by the fisherman by spreading a fine meshed net over the bottom soil of the pond. In shallow ponds these fishes were caught by hand picking.

Poisoning

This method of eradicating the weed and predatory fishes from any sized water bodies is restricted because of high toxicity effect. The selection of poison is carefully done in nurseries and rearing ponds. The poison to be used should have following characteristics.

a) It should be freely available and economical.
b) It must be effective even at low concentration.
c) It should not be harmful to human beings and live stock
d) It should not have any cumulative adverse effect in the future.
e) The dead fish must be usable.

CALCULATION OF DOSAGE

The chemicals are used in ppm and it is necessary to calculate the water volume of the pond, to be treated. Different formulas are used to find out the quantity of water of different sized ponds, for example in square and rectangular shaped ponds this is calculated by multiplying the length and the breadth. In circular pond the area is calculated by Tr^2 and in triangular pond the formula is (Area = 1/2x base X perpendicular to base). The water area is multiplied with average depth to give volume of water. To have more accuracy in the average depth, usually more numbers of possible representative points are taken into account. The quantity of chemicals to be used can be worked out by:

$$\frac{\text{Wt. of water x wt. of application dose of chemicals}}{1{,}000{,}000} = \text{kg. of chemicals.}$$

Chemical Method

Chemicals are used for eradication of unwanted species because dewatering, desilting and drying is difficult in perennial ponds. Plant originated poisons, effects on their respiratory system and the fishes die usually of Asphyxia. The flesh of these fishes may be used for human consumption, but the plant derived poison does not last for long time.

The chemicals used for the fish poisons are:

1. Chlorinated hydrocarbon.
2. Organo Phosphates.
3. Plant originated fish poisons.
4. Application of Ammonia.

Chlorinated Hydrocarbon

The Chlorinated hydrocarbons like, aldrin, dialdrin, endrin and taffidrin-20 are used for the eradication of predatory and weed fishes. The endrin is highly poisonous and effective in killing all fishes at 0.01 ppm and the toxicity lasts for about two weeks. Taffidrin contains 20% endrin as effective poison at 0.01 ppm. These chemicals are mixed in water and sprayed over the pond surface. Fishes die within two hours and float on the surface. The higher dose of taffdrin 0.02 ppm may kill all the aquatic and sub aquatic animals within 4 hours. The common chlorinated hydrocarbons is used for the eradication of fish and predator etc.

These chemicals prepared with water are sprayed over the surface one or two months before stocking the fish.

Organo Phosphates

The common organophosphate compounds used are thiomatin phosphamidon, dimethyl dichloro vinyl phosphate (DDVP) etc. These are used at a concentration of 2-50 ppm. depending on the density of weed fishes and predators. The dose of 0.0032 to 0.5 ppm is effective to the aquatic insects. But organo-phosphates are not suitable for human consumption.

Plant Originated Fish Poisons

There are good number of plants which could be used for fish poisoning. Various parts of plants such as stem, barks, seeds, leaves, roots can be used for fish poisoning. Some of the common plants are:

1. Barring- *Tonia acutangula.*
2. Croton tiglium.
3. Derris root powder.
4. Mahua oil cake.
5. Sugar cane jaggery.

The stem bark, seeds or root of Barringtonia acutangula at the rate range of 10 to 20 ppm are effective. While croton tiglium is effective at the rate range of 4 to 20 ppm. *Millettia piscidia* and *Millettia pachycarpa* are used to eradicate unwanted fishes in Assam.

Derris root powder is commonly used for the eradication of weed fish and other predators. It contains 5% rotinon at the range of 1 to 10 ppm. This poison damages gill membrane causing the death of fish. The effect of the poison lasts for 4-10 days when used at the range of 4-20 ppm. One dose of 4 parts of derris powder in one million part of water is sufficient to paralyses the fishes at 6 to 10 ppm. This powder is used to kill air breathing fishes.

This poison is very effective in shallow ponds to kill all predatory and unwanted fishes. Fish poisoned with derris powder are fit for human consumption within 24 hours. After poisoning, the fishes float on the surface and can be netted out. It these fishes are immediately put in fresh water, the effect of poisoning disappears, and the fish becomes normal. Derris powder is not easily available but can be replaced with other plant derivatives such as Banalu, Akhrot, Rhododendrons, Kuchira and Bantarsbalu are reported to be toxic to the fish. The mahua oil cake used as fish poison as well as fertilizer for ponds. According to Bhatia (1970) Mahua-oil cake was found to be toxic for the fishes at 60 ppm. It contains 4-6% saponin which enters the blood stream of fishes through gills and haemolysis red blood corpuscles causing death of the fish. The mahua oil cake at 200-250 ppm. is reported to kill all varieties of predatory and weed fishes. It acts as manure when its poison effect is lost, it should be applied atleast a month before stocking the carp spawn in nursery ponds.

Sugar cane jaggery is uses at 1% concentration in fish pond for eradication of unwanted aquatics, containing saponin a haemopoietic poison used at 10 ppm. The tea seed cake causes the death of fish within 10 to 11 hours. The seed meal of tamarind is used for eradication of unwanted fishes.

Application of Ammonia

Application of 50% solutions of $Ca(OH)_2$ and $(NH_4)_2SO_4$ in equal quantities is known to release ammonia which kills all predatory fishes including air breathers within 12 hours.

Other Miscellaneous Method

The calcium hypochloride commonly known as bleaching powder contains about 30% chlorine and Urea and bleaching powder is applied for removal of unwanted aquatics and fishes. Even turpentine at the rate of 250 litre/he is applied for the eradication of unwanted fishes. It affects the gills and checks effective gas exchange through gills causing suffocation and eventually death of the organism.

Biological Methods

The biological agents like amphibians are controlled by repeated netting, traps and scoop nets. Frog tadpoles are eradicated by the application of tea seed cake @25ppm. Reptiles are controlled by fencing and scaring. Those inside the pond are removed by repeated netting. Various birds like king fisher, herons and ducks feed on fishes and are removed by scaring, trapping and shooting.

The most important mammal feeding on fishes is Lutro lutro. Other mammals feed on fishes are Beavers and Recons. These mammals can also be controlled by fencing, scaring and shooting.

CONCLUSION

Thus, we can say that the eradication of predatory and weed fishers with the help of all the above mentioned methods results in facilitating carp culture.

Chapter 11

Transportation

INTRODUCTION

Transportation of fish is one of the important aspects of fish culture as both live and dead fishes can be transported from one place to another place as per requirements. There are transportation and storage facilities in certain states of India for example, Tamil Nadu, Bihar, Orissa, Madhya Pradesh, W. Bengal and Maharashtra. It is being much improved as several ice and cold storage plants are constructed at some places in Mumbai, Chennai, Travancore, Cochin, Andhra Pradesh and Orissa etc. Tamil Nadu Government has also set up some freezing and cold storage plants at Kozhikode and Mangalore. Large quantities of fish are transported from the chilka lake area to Calcutta and from Kozhikode to Chenani and Bangalore. Considerable quantities of fish are also moved by fast trains from one area to another. Fish is transported at half the usual rates, but the fish spawn, fry and fingerlings meant for stocking operations, at one fourth the rates. Greater rail facilities is necessary to develop fish industry. During transportation, several factors are taken into consideration *e.g.*, the species of fish, size, water, temperature, length of time taken, atmospheric pressure and length of stops etc. Transport of fry is carried out very carefully. Fish spawn and fry are also transported by aeroplanes from Kolkata to other states, otherwise they will die when released and in that case it will be a waste of efforts, time and money.

IMPORTANT POINTS FOR TRANSPORTATION

1. Reduction in the rates for fish transportation.
2. Movement of larger quantities of fish by fast trains.
3. Introduction of insulated or refrigerated rail transport on all routes connecting important landing centres and large areas of consumption.

FISH SIZE FOR TRANSPORT

The foregoing general factors govern fish mortalities when applied to the transport of spawn, fry and fingerlings separately.

a) The rate of consumption of oxygen in terms of mg; 02/kg body weight is greater for spawn and fry than for later stages.
b) Scales being unformed or in the process of formation, the metabolic exchanges with external medium may be carried on to

some extent by the general body surface also.

c) Inspite of every case, some mortality is likely to take place in spawn and early fry stages.
d) Dead spawn and fry will disintegrate and pollute the water more quickly than bigger fish.
e) They are less prone to suffer hyperactivity injuries and diseases.

In younger stages risk of loss and uncertainly are great than the later stages. The Chinese fishermen experienced that in fish, seed fry smaller than 40 mm. is unsuitable for long distance transport.

CAUSE OF MORTALITY

Mortality of live fish at all stages during and immediately after transport may be caused by.

Oxygen Starvation of Blood and Tissues

The influence of oxygen concentration in transport water depends upon the rate of oxygen consumption. It plays an important role on the mortality of live fish during transport. The rate of oxygen consumption of fish varies considerably depending upon the species, size, state of physical activity, feeding and external factors like water, temperatures, concentration of dissolved oxygen, free carbon dioxide, ammonia, pH level and certain other factors etc. During different types of excretion the oxygen level is reduced in the transport medium on account of bacterial decomposition. If excited, the concentration of lactic acid increases in the tissues and create what is called as "oxygen debt". In younger and smaller fish, the greater is the rate of oxygen consumption per unit body weight. The presence of sufficient dissolved oxygen in the water does not necessarily mean that the blood and tissues are sufficiently charged with oxygen. Low pH, high carbon dioxide and ammonia levels affect the ability of haemoglobin to combine with oxygen. Increase in water temperature accelerates the metabolic rate.

Accumulation of Toxins in the Transport Water

Large quantities of ammonia is excreted by fish and this is supplemented by ammonia produced by bacterial decomposition of urea. Other nitrogenous wastes, dead fry and slime, soon gets built up to a dangerous level. In live fish transport unit, rise of carbon dioxide concentration and lowering of pH are associated factors. Both inhabit the ability of hemoglobin to combine with oxygen. At carbon dioxide concentration of 300-350 ppm. the fish are unable to take up any oxygen regardless of how high the oxygen content of the water is? The rate of oxygen consumption could be reduced by an increase of carbon dioxide concentration within safe limits, to the extent its presence in the water may prove beneficial. Bacteria, plankton and organic matter consume oxygen and produce toxic end products. Better results may, therefore, be expected by packing the fish in clear well aerated bacteria free water. Mortality in case of spawn, once started escalated quickly to a mass scale.

Hyper-Activity Strain and Exhaustion

Fish takes a long time to overcome the "oxygen debt" caused by hyper activity. Many of the surviving fish die during the next few hours, when released in oxygen rich water after handling and transport.

Diseases

Besides external infection it is quite possible that internal infections get easily transmitted among the lot densely crowded in a limited space. Sooner or later, these diseases may become fully manifest and cause mass mortality after packing.

Physical Injuries and Predation

On account of hyper activity, the bigger fish often suffer injuries, which by themselves may cause death or secondarily by serving facilities for external infections.

METHODS OF PACKING AND TRANSPORT

Transport of fishes depend on the size, number locality and weather conditions. The resistant fish can be transported dry in buckets without water or in baskets. Other fishes can be carried in cans, pans, vats and other receptacle carried by hand or vehicles or all sorts like burrows, wheel burrows, motor lorries, different types of vans or trailer, railways and even aeroplanes. Transport by aeroplanes provides excellent service particularly for brood, fish, eggs and fry.

Open System

The open system is without artificial aeration/oxygenation water circulation.

A standard design of open live fish carrier that has been in use with Government Fishery Department is made of galvanized tin, round and square, 40 to 50 litres capacity mounted on a wooden base to prevent heating by contact with the ground. A perforated dish-like insert lid is provided, which serves well for aeration by cascading down the water spilled out into during transport on account of splashing. For transport of fingerlings truck mounted open tanks with facilities for mechanical aeration and on water circulation had been used quite successfully (Hora and Pillay 1962). Open canvas container are already being used in Punjab and Madhya Pradesh for transporting fingerlings. In China, bamboo baskets and wooden tubs are used. In Indonesia water tight, tar-coated, plaited bamboo baskets of 15 litres capacity to accommodate 10,000 fry of 50 mm size are used. In Philippines and Indonesia boats designed to allow free water circulation in the boats, where fish are held is used. A Russian shipyard is building a prototype of a vessel with 3 tonnes hold capacity for transporting live fish with in the country.

Closed System

In this system the source of oxygen supply is not the open air to which the water surface is exposed, but air or oxygen is supplied into an enclosed atmosphere above the water. The oxygen sealed packing method for transport of spawn and fry has gained very quick popularity in the country. In U.S.S.R. carp larvae are transported in special plastic bags filled with water and oxygen in the ratio 1:1 and packing density of 2000-4000 litre.

In India, fry upto 30-40 mm size are transported in this manner for durations of 20-30 hours. In the case of bigger fry the usual packing density is about 7 per litre and mortality less than 5%.

TRANSPORT OF LIVE FISHES

I. Receptacles: The receptacles are adapted not only to the size of the living fish but also to the species and the quantity of fishes which are to be transported from one place to another.

Fish fry and fishes are also transported by bamboo receptacles in the far eastern countries.

Cans

When small quantities of fry or average size fishes are to be transported over a short distance, small can are generally used. These are known as "Boutonnieres". But when large quantities of fry and fish of average size are to be transported flat bottom cans are preferred. Cans are oval or circular with a capacity of 20 to 40 litres. The cans are made of metal or plastic sheets. Zinc cans are lined inside with a protective layer in order to avoid poisoning of the fry. Cans are having two operations which have stoppers. It can be emptied with the help of lower stopper and water can be renewed with the help of upper stopper. A special can is used in mountains for carrying the fishes by a man on his back.

Dry Transport

When the large resistant fishes are to be carried to short distances, baskets and buckets with perforated bottoms are used. There is no need to carry the fishes in water.

Barrels and Vats

Fishes are also carried in barrels and vats. Barrels are old typed and less used. Wooden vats have been replaced by metal or plastic of the same design. These are conveniently carried by vans. Wooden barrels have solid handles on either side. The shape of the barrel is such that they facilitate oxygenation and automatic aeration goes on when they are carried. The opening of barrel is quite large. The covers are truncated pyramid like. They are hollow to enclose ice. The bottom is pierced with holes for flowing water from the melting ice.

Vats are like truncated cans or cylinders resting on flat end with round or oval circumference. Vats are of different capacities containing a bottle of compressed oxygen (3-7 litres at 130 to 150 atmosphere) inserted vertically, fixed with collars. Sometime the bottle can also be fixed outside the vat. A pressure reducing valve with a tube carries the oxygen to a spray which is well protected by a cover. The hundies, hoops of barrels and vats are to be protected against rust.

Polythene Bags

Polythene bags are extensively used for carrying small quantities overland or aeroplanes. Generally one third of this bag is filled with water and the fishes are kept in the bag. The oxygen is filled into the bag with the help of a rubber pipe or a bottle. The upper part of the bag is twisted and closed with a solid elastic band, one or more bags are placed in a carton and then are transported. The oxygen above the water diffuses slowly during transport and a balanced oxygenation is ensured.

Special Vehicles

For transporting fishes in large quantities fish farms have motor Lorries, mounted with metallic tanks of different shapes. Tanks are provided with an aeration system. The oxygen bottles or compressed air are in use. There is installation of spray at the bottom of these tanks which ensures the distribution of gas. Sprays are in hard porous materials or comprise a simple metal frame at the bottom of the tank. Air pumps are used which function as the battery system. Some vehicles are equipped with hydraulic pumps, sucking water from the bottom to the top and spraying the same in atomized from the top. Sometimes the pumps are provided with filter to clean the water. Large quantities of fishes are also transported by railways.

Other Methods

Traditional Method

In this method earthen pots of about 15 litres capacity are used for the transportation of seed fish. The pot is first filled about two-third with water of the spawning ground. The pH and dissolved oxygen of the water is to ensure maximum survival rate of seed fish that should be between 7.5 to 8.5 and 4.5 ppm respectively.

After filling the pot with water about 50,000 spawns are introduced. It is advisable to condition the spawns in the hapa for three days without feeding, prior to transportation. The excreta may pollute the water in the earthen pot leading to death of seed fish.

Earthen pots are carried manually on bamboo slings. To avoid the mortality of seed fish due to asphyxiation, water is changed once in every five hours, depending however, on the number of spawn in the pot and the duration of transport.

Transportation in Closed Containers

The seed fish are transported by road, rail or air in plastic bags filled with water and compressed air packed in metal containers. Metal containers made of aluminium, galvanized iron and tin are mainly used.

Tins of 40x25 cm. could be conveniently employed, Polythene bags measuring 65 x 45 cm. are checked first by inflating the bag with little air and immersing them in a container with water. Leaking or damaged bag produces air bubbles in water. Every bag thus checked is kept in a tin container. About one third of its capacity is filled with aerated pond water.

The fry or fingerlings, starved for one day are acclimated and then carefully introduced in requisite numbers into the bag. A tube from the oxygen cylinder is then allowed into the bag and the portion of the bag about 10 cm. from the top is twisted and a string is kept ready for tying. The oxygen is then drawn in from the cylinder through the tube, until two-third of the bag is inflated or the top of the inflated bag is slightly below the top of the tin. As soon as the inflated condition is obtained, the string is tied round and the tin is closed. Such packed tins are kept in a cool place until they are transported to fish culture sites. To ensure better survival rate the tins should be transported during evening or early morning. If the culture sites are distant, the tins in transit may be kept, cooled by covering with water soaked gunny bags. Recent investigation on carps shows that by anesthetizing the seed fish they could survive for longer period by minimizing the concentration of toxic gases like ammonia and carbon dioxide.

General rules for Transportation

For the transport of fish and fry in particular, supply of sufficient oxygen in the receptacle should be ensured. For this purpose water must be kept at a low temperature and there should be arrangement to change (renew) water at different stops. Movements of receptacle should be moderate. There should be adequate aeration by supply of air or oxygen.

There are some important factors which influence transport of living fishes.

The Species of Fish

The requirements of oxygen vary considerably accordingly to species of fishes. For salmonids there is sufficient need of oxygen.

Age and Size of the Fish

The requirement of breathing is higher not only in the large fishes but the small fishes also require greater amount of oxygen corresponding to their total body weight.

Length of Time of Transport

When the time taken is comparatively short the stocking rate is greater.

The Temperature of Water

The low temperature is more suitable during transportation as oxygen content at low temperature is higher. The range of temperature is 4 to 10°C which is most suitable for transportation.

Relative Resistance of the Fish

Fishes depending on artificial food are less resistant in comparison to fishes having natural food. Spawning fishes face difficulty during transportation.

The Nature of the Transport Receptacles

The wooden transport receptacle warm up more slowly in comparison to metal. However, the latter can be thermally insulated.

Means of Transport and Lengths of Stops

Faster transport and shorter stops, provide a better chance of success in transportation.

Climatic Conditions

The oxygen content are greatly influenced by climatic conditions. To reduce the metabolism of the fishes and their oxygen intake, tranquillizers have been used.

Ice and Cold Storage

The ice and cold storage facilities are equally necessary in development programmes in Mumbai. The Central Government has set up a large ice factory, quick freezing and cold storage plant in co-operation with their deep sea fishing station. This plant provides ice freezing and storage facilities not only for the catches of the vessels of the Central Government operating at Mumbai, but also to the fishing industry. There are, however, several other private ice and cold storage plants which help in meeting the needs of the fishing industry in Mumbai. TamilNadu Government has set up small freezing and cold storage plants at Kozhikode and Mangalore. Ice and Cold storage plants also exist in some other big towns, but refrigeration facilities are too inadequate in most of the fish landing centers. Several small ice and cold storage plants setup in India under the Indo-American technical co-operation programme, are under construction in important fishing centers in different state *i.e.* Saurashtra, Mumbai, Chennai, Travancore, Cochin, Andhra and Orissa. Some plants are expected to be operated by co-operative societies.

Transport of Breeders

Breeders have to be transported without shock and injury. This may be done using open canvas carries (1xl.25m) or in splash less closed and foam-lined containers with compressed air. It is always better to give dip baths to the breeders in any of the antiseptics or antibiotics, such as methylene blue (2ppm), acriflarine (10 ppm), Copper sulphate (0.5 ppm), Potassium permanganate (3ppm), Chloromycetin (8-10ppm) or Sodium Chloride (3 percent), so as to protect them against infections, such as bacteria, fungi etc. Further, before transport, the breeders have to be tranquilized using any one of the following anesthetics.

1. Quinaldine (2-methyl, quinoline), (0.01-0.05%) 2. MS222 (Tricaine methanesulfonate or m-amino benzoate methane sulphonate –0.1 ppm) 3. Sodium amytal (50-170 ppm) 4. Tertiary amyl alcohol (0.5 percent) 5. Phenoxy ethanol (0.01-0.05 percent).

Transportations of Seed

The transportation of fish seed from collection site to the farming site is done with a lot of care. The mortality which is expected during transport owing to following reason prevented by adopting various measures.

i) Mortality may be due to depletion of oxygen.
ii) Due to accumulation of toxic gases like CO_2 and ammonia in the medium of fish seed carrier.
iii) Length of transportation time.
iv) Ambient temperature of medium, carrying the seed.
v) Accumulation of excretory wastes.

The fish seed is transported by closed containers.

Transportation of Spawn and Fry

Spawn and fry are now transported in sealed metal containers with oxygen. The container made of aluminum or galvanized iron or polythene bags are used for this purpose. Kerosene tins fitted with air tight caps and with tubes for drawing in oxygen and letting out displaced water have also been used successfully. Alkathene bags of different size are now widely used in India for transportation of fish seed.

Transportation of Fry and Fingerlings

Transportation of fry and fingerlings in open earthen vessels called "hundies" has been the traditional method in Bengal. During the recent years aluminium vessels are also used. Hundies are of various sizes. Smaller ones re for the transportation by the road and by rail. The hundies are filled with water from the same source and the fry 50,000 and 75,000 are released in vessels of 20-25 litre and 30-35 litre respectively. The small quantity of alkaline soil is added in each vessel, probably to buffer pH fluctuations. The vessels are periodically shaken during transit. For aeration, water is partially renewed at intervals, fry coming up and gasping, is a sign of oxygen deficiency and requires the change of water. Dead fry are removed immediately is avoid pollution.

Large sized open metal containers, with wide mouth can be closed by perforated lids, are now preferred in place of earthen hundies. Aeration of water in the tins containing fingerlings has been tried. Vessels are transported in open motor vans fitted with hand pumps which circulate the water in the form of spray. By this method fry can be transported upto 500 km, with only 5% mortality. Open canvas containers are used in Madhya Pradesh and Punjab for transporting fry and fingerling to carp breeders.

Transportation of Eggs

The fertilized eggs and 24 hours hardened eggs of cold water fish, especially Mahsheer could be transported by packing them in moist cotton wool and placed in plastic bucket with an outer container. However, for transshipping trout eggs, card board cartons of 20 x 30 x 20 cm. size with inner styrofoam lining are used. Inside these boxes, porous polyethylene bags containing about 4,000 eggs are sandwiched between upper and lower moist sponge or cotton pad. Above the upper pad a polyethylene bag containing about 1 kg. broken ice is kept for maintaining low temperature.

Transport of Fingerlings and Brood Fish

In India, two successful models of closed system live fish carrier tanks have been designed that is known as splash less tank. The latest model of the splash less tank is of a petrol tanker design of 1,150 1bs capacity with an autoclave of type lid. It has a built in belt driven by the engine of the transporting vehicle. The oxygen cylinder is carried only as a standby for emergency. The inside of the tank is lined with U-foam which prevents physical injury to live fish during transport. About 250 kg of the live fishes can be transported at a time in the splash less tank. Adult catla 60 kg. is transported in these tanks and also 90,000 carp fingerlings.

The other live fish carrier designed in India by Patro (1968). This carrier is of laboratory gas supply design type and comprises of an outer chamber of 120cm. diameter open from top and a slightly smaller inner one closed from top, the later during transport fits inside the former. The top of the inner chamber is provided with an air vent and an oxygen valve. The outer chamber serves as a storage tanks and is initially filled with water along with fishes to be transported. The inner chamber which is slipped from the upper open end of the outer serves as an oxygen holding chamber at its top and is lined throughout with U-foam, to prevent fish from sustaining injury during transport. This double barrel type carrier as named by Patro can transport a total weight of 100 kg of live fish at a time.

Transport of Freshly Killed Fish

Even the dead fishes should be transported at a low temperature to avoid the danger of decomposition of their flesh, the temperature should be between 0° and 4°C and as such the fishes are placed in crushed ice. This method stops the development of bacteria and also enzymatic and oxidative changes.

Use of Chemicals in Live Fish Transport

The sedatives and disinfectants are also used in transporting medium. The sedation of fish brings in practical benefits by way of –

1. Decreasing the rate of oxygen consumption and reducing the rate of excretion of carbon dioxide, ammonia and other toxic wastes.
2. Controlling the excitability of the fish and thereby reducing chances of injury.
3. Reducing the time required for handling them. The behavioral responses that occur in fishes during anaesthesia as given by McFarland 1960.

For Example: For fishes the behavioral responses, the reaction to external stimuli; equilibrium and muscle tone is normal, but light sedation the fish slight loss in reactivity to external stimuli loss of reflex activity the total loss of reactivity respiration rate very slow and heart rate become slow.

McFarland recommended a dose of sodium amytal in the range 0.2 to 0.65 g per gallon (52 to 172 ppm) of water depending on its salinity and the effect desired. Sodium amytal is antagonistic to calcium and can therefore be used only in soft water. The anesthetics chloral hydrate, tertiary amyl alcohol and methyl paraphynol (Dormison) are, however, considered more desirable by McFarland (1960). These narcotics have intermediate potency, are readily soluble in both fresh and saltwater. Induce deep sedation within 10 minutes are compatible with calcium. The concentration recommended for deep sedation with these anesthetics are tertiary amyl alcohol 2ml/gallon, dormison 1-2 ml/gallon and chloral hydrate 3 to 3.5 g/gallon.

Table 11.1: Behavioural responses of fishes in relation to different levels of Anaesthesia (Mcfarland, 1960)*

Definite levels of anaesthesia	Behavioral response of fish
Stage	
0 Normal	Reactive to external stimuli, equilibrium and muscle tone normal.
I (1) Light	Slight loss in reactivity to external
I (2) Deep	Total loss in reactivity to external
Sedation.	Stimuli except strong pressure, slight decrease in opercular rate.
II (I) Partial loss of equilibrium.	Partial loss of muscle tone, react only to strong tactile and vibrational stimuli, rheotaxis present, but swimming capabilities seriously disrupted, increased opercular rate.
II (2) Total loss of equilibrium.	Total loss of muscle tone, react only to deep pressure stimuli, decreased opercular rate below normal.
III Loss of reflex activity.	Total loss of reactivity, respiratory rate very slow, heart rate slow.
IV Medullary collapse	Respiratory movements cease, followed several minutes later by cardiac arrest.

MEASURE FOR SAFE TRANSPORT

1. The capture and handling of fish should be done when the habitat is the coolest *i.e.* during early hours of morning
2. The fish should be segregated size wise and if possible species wise.
3. Warming water should be avoided. For this receptacles should not be placed in direct rays of the sun.
4. In a hot weather the cans or barrels should be kept in contact with the ice blocks. Ice is never placed in the water directly.
5. Cans can be covered with cloth, soaked with water, thus evaporation of this cloth can maintain freshness to some extent inside the cans.
6. The fry and fingerlings should be conditioned *i.e.* held in concentration in clear habitat water for a duration of a few hours

to a few days depending upon the size of the fish.

7. Frequently change of water over a long journey, entire water should not be changed, as it will lead to temperature difference and harmful for the fishes. Water should be changed slowly.
8. Fresh water should be unpolluted and well aerated. Sometimes air is pumped into the water for aeration when urgently needed.
9. Optimum load density must be decided case by case based on results of standing tests. Oxygen volume should be strictly need based so that greater volume of water could be used to mitigate the concentration of toxic pollution and accommodate more fish.
10. The fishes against the side of the carrier get shocks of that will injure the fishes and the latter will gradually die, so, jerking should be avoided.
11. Spawn and fry should be packed in smaller units.
12. The carrier should be painted white to reflect radiant heat with a lining of thermal insulation.
13. A transparent leak proof observation window should be provided.
14. The fishes are transferred to a new environment, they must be acclimatized slowly and progressively.
15. Quantities of fish should vary according to type of transport *i.e.* with diffused oxygen or without diffused oxygen. Excessive oxygen can be harmful for the fishes particularly to the health of the fry.

CONCLUSION

Transportation of fishes to various places provides a proteinous food for human being and in a way solve the food problem of world.

Chapter 12

Sewage Fed Fisheries

INTRODUCTION

The Term "sewage" broadly refers to all the liquid discharge from domestic and industrial sources. It contains about 99% of water and one percent of other materials including bacteria pathogens, various organic and inorganic substances present in true molecular, colloidal or semi colloidal state. The sewage differs from sludge which consists of liquid wastes from kitchen and bathroom, but excludes faecal matter and urine. In many countries, sewage is treated and used for drinking, industrial and agricultural purposes (such as in culturing fish and field crops). Fish is known to convert bio-waste into edible proteins. It is widely used, to increase fish yield, in the countries like Germany, Poland, Hungary, Israel, China, Indonesia, Japan etc. using domestic sewage which serves as a source of fertilizer and food for fish.

Use of sewage in fish culture is a profitable proposition because it is not only a cheap source of organic manure to the fish pond, but also consumed as food by fishes. Initially the chief fisheries were located in West Bengal and Chennai but the use of scientific methods of sewage treatment and its application is profitable in fish raising and is realized in all the states of the country.

The fish seed farm at Durg in Madhya Pradesh for example has principal source of water in two sewage stabilization ponds and gives annual income of Rs. 3 lakhs annually. Vidyadhar sewage spill area in Kolkata gives an average yield of 1200 kg of crops per hectare area. In coastal areas the sedimented sewage is diluted with brackish water supplied to the paddy fields for profitable fish yield.

Characteristics of Sewage

The raw sewage is, however not supplied to the fish pond because of the following characteristics.

1. High bio-chemical oxygen demand (BOD).
2. Low dissolved oxygen (DO) contents.
3. The bacterial load.
4. High carbon di-oxide contents.
5. High values of ammonia and hydrogen sulphide.
6. If sewage is properly treated it is of considerable value in raising the fertility of pond water.

7. The cloudy fluid contain organic and inorganic matter.

Quality of Sewage

In domestic sewage water content is about 99%. The rest consists of organic and in-organic constituents. The common gases and salts present in sewage are carbon dioxide, hydrogen sulphide, ammonia, nitrate and phosphate. The other heavy metals present in traces are zinc, copper, chromium, manganese nickel and lead. The dissolved oxygen content of sewage is nil. Apart from the abiotic components the sewage harbours a variety of bacteria and protozoans. The sewage is beneficial for fish culture because.

1. The smaller particles of sewage form direct food source to zooplankton and benthos.
2. The larger particles of sewage are directly utilized by fish.
3. Soluble organic substances are directly used by phytoplankton for photosynthesis and by zooplankton as a food source.

Composition of Sewage

According to Saha *et al.* (1958) sewage of Kolkata has the following composition.

1	Dissolved oxygen.	Nil
2	Dissolved Carbon dioxide	20.96 ppm
3	H_2S	2.4—4.8 ppm
4	Free Ammonia	12.0-63.6 ppm
5	Nitrite	0.0-0.08 ppm
6	Alkalinity	170-490 ppm
7	Phosphate	0.01-0.33 ppm
8	Suspended Matter	160-420 ppm
9	pH	6.9-7.3 ppm
10	Water	99%

Composition of Indian Sewage

The average composition of Indian sewage is:

a) 5 days BOD of 280 mg. Litre.
b) 402 mg/litre of suspended solids.
c) 1000 mg/litre of dissolved solids.
d) 200 mg/litre of chlorine.
e) 25 mg/litre of ammoniacal nitrogen.

The point of interest is that, the rise in temperature increases the value of 5 days BOD.

Types of Sewage

Domestic Sewage

It is the liquid waste discharged from the kitchen, bathroom (faecal matter and urine) having carbon, nitrogen, ammonia, phosphorus, bacteria, protozoa and 99% of water.

Industrial Sewage

The effluents released from the different industries containing large amount of inorganic and organic compounds when treated with different chemicals, certain compounds either precipitate or coagulate and some by sedimentation settle down, then it becomes usable.

THE OBJECTIVE AND NEED OF TREATMENT

The principal objective of treating sewage is two folds to reduce toxicity and possibility of spread of diseases by pathogens and to prevent pollution of surface and ground water. One of the ideal ways to prevent such pollution of surface waters is to make beneficial and use after pretreatment *e.g.* by growth of algae and fish by irritation of soil.

This is followed by degradation or disintegration of organic matter into easily assimilable form with liberation of methane by biological organisms and finally refining the sewage so as to make it fit for reuse.

TREATMENT OF SEWAGE

The process commonly applied in sewage treatment according to Imhoff *et al.* (1956) essentially consists of.

1. Coarse suspended solids: Screening.
2. Heavy settleable solids: grit chamber detritus tank,
3. Floating solids including fats and oils: Skimming using grease traps or skimming tanks.
4. Fine suspended solids: flotation, skimming and sedimentation, chemical precipitation and fine screening mechanical filtration.
5. For highly dispersed organic solids, including dissolved solids: biological treatment, including treatment by sewage irrigation, Intermittent sand filters, trickling or percolating filters, sewage ponds, septic tanks or anaerobic ponds, activated sludge works.
6. Deodorization: Chemical treatment, using chlorine, ferric chloride etc.
7. Disinfection or sterilization: Chemical treatment using chlorine, copper sulphate, heat treatment etc.

Other methods of sewage treatment are:

1. Mechanical Methods

The mechanical process consists of screening and filtration, so as to remove coarse suspended matter. Floating solids including fats and oils as well as fine suspended material are removed, by skimming and sedimentation. Sedimentation is done by letting the sewage into a tank at a high velocity, when the sewage enters a large tank from a sewage channel, there is a sudden drop of velocity resulting in sedimentation.

2. Chemical Method

This chemical method includes coagulation precipitation, deodorization using chlorine, ferric chloride, lime alum etc. and disinfection by using chlorine and copper sulphate.

3. Biological Method

The biological treatment of sewage consists of bacterial decomposition of organic contents of sewage to carbon-dioxide, nitrogen oxide, sulphate etc. The bacteria decomposes the substances either aerobically or anaerobically.

Arceivala et al. (1970) used this term for both aerobic and anaerobic stabilization including both sewage and industrial wastes. This term is applied to a water body, artificial or natural, retaining sewage or organic waste water until the waste are rendered stable and inoffensive for discharge into receiving waters. Waste stabilization ponds are especially suitable in India because of plentiful sunshine.

They are cheap and easy in operation. Organic matter contained in the waste is stabilized and converted in the form of algal cells which go alongwith effluent. They are classified according to the biological process taking place in them as.

(1) Aerobic (2) Anaerobic and (3) Facultative Ponds.

Aerobic Ponds

They are shallow 0.3 m in depth. They are designed so that algal growth through photosynthetic action is maximized and waste material is stabilized through microorganism.

Anaerobic Ponds

They are 2.5 to 3.7 m deep. They take high organic loadings so that anaerobic condition is maintained throughout the pond.

Facultative Pond

They are 0.9-1.5 m deep and are aerobic throughout the day as well as few hours in the night, but only for remaining hours of night the water layers become anaerobic. In India many of the stabilization pond are a combination of aerobic and anaerobic type. (Fig. 12.1).

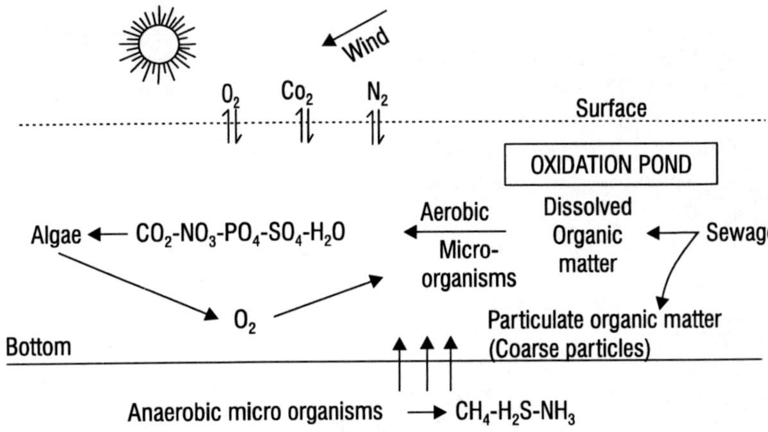

Fig 12.1: Processes involved in sewage treatment

General Method

For fish culture sewage is treated by sedimentation dilution and storage.

a) **Sedimentation:** The process is completed in two sedimentation tanks of which the first is built at a higher level than second. In the first tank the sewage is retained for 10 days so that the organic matter is decomposed aerobically into nitrates, phosphates, sulphur etc. The second tank known as waste stabilization tank, receives sewage slowly that falls at high velocity into it. Sudden drop in velocity of sewage in the second tank causes further sedimentation and homogenization of sewage. Variants sewage is allowed to stay for 15 to 20 days, during which water loses, its foul odour and becomes green due to algal blooms. The sedimentation tank (ST) are of three types viz. the aerobic, facultative and anaerobic. The aerobic matter in aerobic sedimentation tank (.5m deep) decompose organic matter aerobically into stable ones. Consequently algae grow to the maximum extent. Facultative Sedimentation tank (.9 to 1.5m deep) is aerobic in the day and becomes anaerobic in the night, Anaerobic ST (2.5 to 3.7 m deep) hydrolyze and ferment organic substances into volatile acids, alcohols, CO_2 and methane by means of two different kinds of bacteria.

b) Dilution: The stabilized and odorless sewage which flow out of second tank is diluted with fresh water in the ratio of 1:5 (one part sewage and five parts of water) to rest or dissolved oxygen and reduce concentration of CO_2, NH_3, H_2S etc. It can be further improved by photosynthetic activity of algae. This diluted sewage is supplied to the nursery or rearing ponds where carps feed on algal blooms and attain large size.

c) **Storage:** The sewage water added to stocking pond is beneficial for fish growth as it is rich in fertilizers and food, and the biological process carried out by microorganisms present in the raw sewage, oxidize it. (Fig. 12.2).

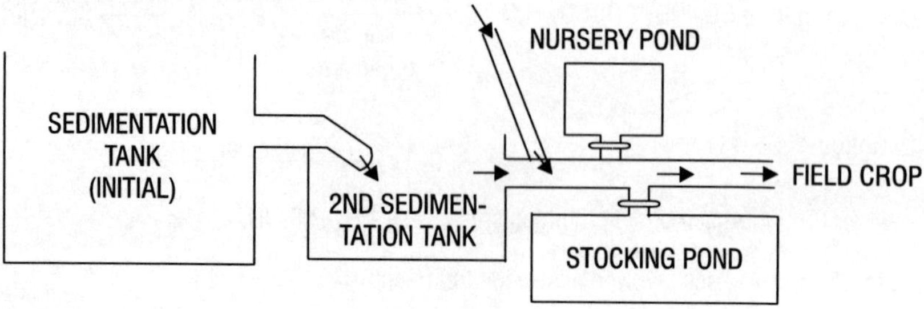

Fig 12.2: Showing Plan of Sewage to Fish Ponds

Description of Oxidation Ponds

The oxidation and fish ponds should be located in open space, so the sewage is treated in shorter time. The sewage may be conveyed to oxidation pond by gravitation to discharge sewage at the bottom of the oxidation pond. The oxidation pond should be located on high elevation than the fish pond. The size of oxidation tank varies. It is selected according to input and out flow of sewage from it, keeping in view that the sewage can be discharged from it for 10 days continuously. If the daily in flow of one lakh litre of sewage, a pond of 50x20x15 mtr. is suggested, one mtrs. depth may be enough for storing the sewage and 0.5m. free board of space above the water column is essential.

City sewage is used for fish culture in Kolkata. The sewage is diverted through a system of drains to sedimentation tank. After sedimentation and dilution at 1:4 with fresh water, it is added to fishery ponds at monthly interval. Fingerlings of rohu catla and mrigal are stocked in the ratio of 1:2:1 giving an average yield of 1,850 kg/hec. of fish.

EFFECT OF EFFLUENT ON NATURAL SOURCES OF WATER

Pollution of natural water by sewage and industrial wastes is objectionable and damaging for many reasons. Of primary importance is the possible hazards to public health and safety. Of lesser consequence but real are change to streams and destruction of the economic value of clean natural waters. Considering all these ill aspects of water pollution, different pollution control agencies in the world have proposed limits in relation to various pollution parameters of different uses.

Sewage Fed Fish Ponds

The number of fish ponds should be calculated in such a way that their total area is equal to that of oxidation pond. All species suitable for fresh water fish culture are also suitable for sewage fed fish culture. However, the Indian major carps *viz.*, catla-catla, rohu and mrigal are more suitable considering the rich nutrient status of the treated sewage and high survival rate (70-80%) of fingerlings in sewage. The stocking density of carp fingerlings can be 20,000/hac (catla 5,000 rohu 10,000 and mrigal 5,000 in number) A total fish production of 6/hec has been achieved by sewage fish culture. After natively the Indian major carps, silver carps, grass carp, common carp and fringe lipped carp may also be stocked in sewage fish pond at a stocking density of 10000/hac as shown in the given table. It may be observed that such a fish yield results without the addition of fertilizers and supplemental feeds.

Further, the growth rate of fish has also been found to be high in sewage system. It is recently found that high yield of giant fresh water prawn *Macrobrachium rosenbergii* could be obtained from rice fields, supplied with domestic sewage. Air breathings and low oxygen tolerating fish like murrels and cat fish may also be grown along with the species already mentioned, while they should be introduced only after the fingerlings of other fish have grown to a considerable size. In oxidation ponds murrels can be directly cultured and the rapid breeding tilapia, the youngs of which form food for murrels.

Table 12.1: Stocking ratio for sewage fed fish pond.

Variety of fish	6 variety stocking (%)	5 variety stocking (%)
Silver Carp.	25	----
Catla	15	40
Rohu	10	10
Mrigal	25	20
Common Carp	20	20
Grass Carp	05	---
Fringe-lipped carp	---	10

Aesthetic considerations may often limit the use of sewage cultured fish. These can be overcome by transferring the harvested live fish to fresh water ponds and keeping them for a couple of weeks when the objectionable odours and pathogens, if any, will be removed making the fish marketable. Bacteriological tests conducted on the skin, gut and muscle of fishes grown in sewage and fresh water showed that species of salmonella were not present in any of the fishes. Shigella was isolated from fishes treated in sewage fed ponds. But neither salmonella nor Shigella was present in cooked fish.

Sewage Water for other Crops

The drained sewage could be used for irrigation especially for fodder crops, vegetable crops like coconut, palm etc.

Modern Method (Model Plan)

Depending upon the amount of sewage discharge, the size and number of fish ponds may be decided. The plans of sewage fed fish culture in using suitable combination of fish species and fish culture unit suitable for a civic body and educational institution with 500 residents. (Fig. 12.3).

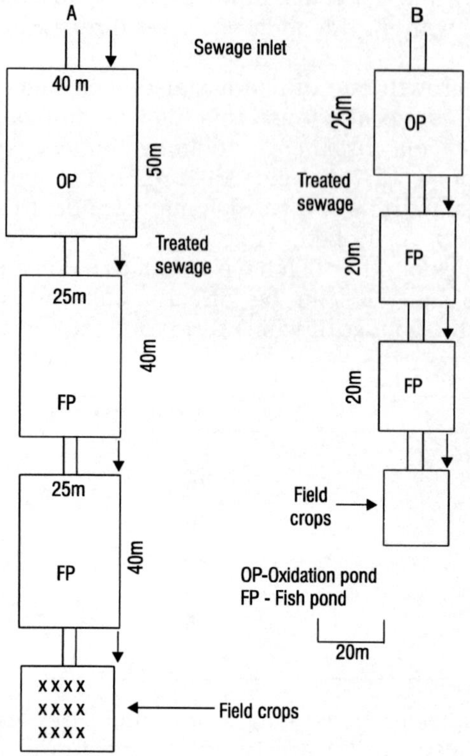

Fig 12.3: Plans for sewage-fed fish culture (A) suitable to civic body (B) educational Institution

RECENT RESEARCH FOR SEWAGE FISH CULTURE

During recent years interest in fish culture in sewage oxidation ponds has increased considerably due to its being more profitable. In a recent experiment (Akolkar and Belsare 1984) fingerlings of common carp Cyprinus carpio were stocked in the primary oxidation ponds at Bhopal at the rate of 80 kg/hec. (approx. 5000) fishes were netted periodically, after three months. There was increase in body weight every month, and the fish reached 1197 gm. within one year. The total average production was 11,938 kg/hec. year. Analysis of the muscle of the fish showed 22% protein 3.2% fat and 10.4% of glycogen. Thus, these fishes had more nutritive value than those reared in other ponds.

Ghosh *et al.* (1976) reported a total production of 220 kg/hec. area in a poly culture of tilapia and Clarias. In other experiment (1979) 710.7 kg of tilapia fishes were harvested in 0.076 hec. of sewage irrigated pond, corresponding to a production of 9.350 kg/hec. annum. It is evident that the culture of tilapia in both the mono or poly culture system gives best cost effect of their raising.

The stocking densities for catla, rohu and mrigal stocked together in treated sewage fed water is suggested to be 20,000 per hectare in a ratio of 1:2:1. Stocking with other carps is suggested to be 10,000 per hectare in the following ratio.

1. Silver carp, catla, rohu, mrigal, common carp and grass carp in a ratio of 25:15:10:25:20:5.
2. Catla, rohu, mrigal, common carp and fringe lipped carp in ratio of 40:10:20:20:10.

Cultivated Fishes

Fishes like *Clarias batrachus*, *Heteropneustes fossilis*, Channa, Tilapia mossambicus and Ctenopharyngodon idella can survive and grow in these ponds which has poor dissolved oxygen contents. The most suited for sewage pond culture are the species of Tilapia, which demand lesser dissolved oxygen, having fast growth rate and breed easily. They can survive at high ammoniacal nitrogen level of 5.43 ppm. These fishes mature at a small size and begin to reproduce earlier before they grow in size. To control the population of fry, cat fishes are introduced. Ghosh et al (1976) reported total production of 220 kg/hec. area in a polyculture of tilapia and clarias. In an experiment (Ghosh *et al.*, 1979) 710.6 kg of Tilapia fishes were harvested in 0.076 hec. of sewage irrigated ponds. Carps are sensitive to dissolved oxygen content of water and thus raised in pond diluted sewage water. A total production of 6 ton carps per hec. area has been produced but the fingerlings of Cyprinus carpio stocked at the rate of 80kg/hec. gave an annual yield of 119.38 kg/hec.

ECONOMICS OF SEWAGE FED FISH CULTURE

The recurring expenditure on sewage fed fish culture is meagre compared to that of fresh water fish culture. The economics of sewage fed fish culture is that this type of culture is lucrative and a fish farmer can obtain an income on an average of Rs. 41,000/hec/yr. If murrels are cultured in oxidation ponds and the excess sewage is utilized for the cultivation of crops, the revenue could be further augmented. (See table).

Economics of Sewage fed fish culture under taken in a 0.1 hec/area.

Operational Cost	
Cost of 1,000 fingerlings (Stocking density, 10,000/hec).	Rs. 200/-
Miscellaneous Expenses.	Rs. 200/-
Interest at the rate of 10% on	
Initial investment.	Rs. 300/-
(Value of 1ha pond.	Rs. 30000/-
Income	
Value of fish harvested at Rs. 8/- kg.	
(Average growth of an individual fish 750gm. and survival 80% the yield would be 600kg/ 0.1 hec.	Rs. 4800/-
Costs and returns.	Rs. 4800/-
Income	Rs. 700/-
Total Cost	
Net Income/0.1/hec/yr.	Rs. 4100/-
Estimated. Income/hec/yr.	Rs. 41000/-

EFFECT ON PUBLIC HEALTH AND SAFETY

The principal hazards to public health from polluted streams are mainly due to the presence of pathogenic bacteria from domestic sewage, toxic materials (acids, alkalies, specific poison like cyanide and heavy metal salts) and materials from industrial wastes. Danger to public health from polluted waters is also from consuming fish food that have lived in the polluted stream.

EFFECT OF AQUATIC AND OTHER LIFE

Pollution of natural waters may be disastrous to fish and other organisms naturally inhabiting the stream. Organic wastes under the influence of bacterial action deplete the dissolve oxygen by consuming it in biochemical oxidation reaction. This may suffocate the fish and other aquatic animals. With mild pollution fish may acquire a flavour that renders their flesh unfit for use as food, whereas with more severe contamination the fish become sick or die.

DAMAGE TO PROPERTY

Discharge of acids and occasionally other industrial wastes leads to the damage of property through corrosive attack, corrosion of steel objects pipe, pumps, valves and other equipments in plants using contaminated water as cooling water, are quite common.

WASTE WATER

If untreated waste water is allowed to enter a river, certain problems will arise.

1. Spread of disease caused by bacteria, virus and parasitic organisms.
2. Oxygen will be reduced in rivers and lakes which receive waste water.
3. Poisoning of the lakes and rivers by chemicals from industry.
4. Enrichment of natural waters by thus increasing algal growth.

Thus oxygen diffusion rendering unsuitable for aquatic life if no further organically rich effluents enter the river, it will gradually recover its oxygen content and will return to normal form.

CONCLUSION

It is evident that the sewage fish culture is profitable, because it does not require fertilizers and supplemented food, thus reducing the cost of production. The growth of these fishes is relatively higher. Recently high yield of *Macrobrachium rosenbergii* (fresh water prawn) is recorded from the paddy fields supplied with the sewage water.

Chapter 13

Fecundity

INTRODUCTION

The reproduction linked with the life cycle of ensuring the propagation of its own race. The size quality and abundance of the spawning populations, along with the condition in which eggs and larvae developed are also taken into consideration. It is connected with the mode of life of an adult fish.

The reproduction in fishes includes number of features which are specific to aquatic animals. In fishes, fertilization of ova takes place in water. Eggs and sperms are laid in water. The fecundity of fish is much higher than the terrestrial vertebrates. The most fecund fishes have floating as well as pelagic eggs from which spawns emerges to attach themselves on the leaves and stems of water plants. Fishes having protective devices for the eggs which are usually provided with low fecundity. The fecundity in marine fishes is usually higher than that of fresh water. In migratory fishes the number of eggs contained in an ovary of a fish is termed as the individual total fecundity. The size of the eggs varies. The pelagic eggs are about 1mm. in diameter, but some are even smaller. The eggs of the fresh water fishes are usually much larger, the eggs of Bitterling, Rhodes are 3mm and Siluroid have eggs 17-18mm. The largest egg is of the elasmobranches, sharks and rays which are laid in a hard shell or capsule. The eggs of most fishes have shells. The egg shell is a tender membrane when ready to leave the ovary and gets covered by shell secreted by glands. In many skates and some sharks, the shells are horny, hard and almost nail like in strength and resist damage.

Fish eggs have many other characteristics, some are buoyant and have a specific gravity like of fresh water ova. Most eggs, are adhesive and sticky while some eggs are non-adhesive, others that sink, however, are at least temporarily adhesive.

The number of eggs that are produced by an individual female varies greatly accounting to different factors including its, age, size, condition and species. The number of course also varies somewhat according to the size of the eggs in relation to the size of the female and the number of eggs in the ovaries of an individual that are ripening for the coming spawning period. This factor is useful for racial studies in taxonomy and for studies on population and production.

The fecundity difference can be observed in fishes of the same length if kept under different conditions. The substantial differences are observed in the fecundity between closely related species and sub-species.

DEFINITIONS

The capacity of an individual fish to produce female reproductive cells is known as fecundity.

Welcomme (1967) used fecundity for the number of eggs produced by a mouth breeding cichlid and fertility for the number of young's produced.

Raitt and Hall (1967) use three definitions with the viviparous red fish Sebastes marinus as.

1. Pre fertilized fecundity *i.e.* number of eggs before fertilization in an ovary.
2. Fertilized fecundity *i.e.* number of fertilized eggs in an ovary.
3. Larval fecundity *i.e.* number of larvae after hatching before extrusion.

Gerking (1967) suggests that in a mouth brooding Sarotherodon, can be conveniently categories as ovarian fecundity and Brooding fecundity.

Identification of Eggs

Spawn collected from nature requires the identification of eggs, such parameters are important at the time of collection. For eg. Collection diameter of egg, shape of egg, colour of egg, size of Peri vitelline space distribution and nature of oil droplets etc. In the examination of the embryonic stages in the eggs, they are treated with a mixture of glacial acetic acid (approx. 3-5 ml) and 0.7% sodium chloride solution (100 ml). This makes the shell transparent and embryos appearing white and the coloured yolk.

These eggs are preserved by keeping them in 10% formalin containing glacial acetic acid (2-5%) or Hydrogen peroxide (2-5%).

Categories of Eggs in the Ovary of Fish

1. The recruitment stock of eggs consisting of rather small, white and opaque looking eggs.
2. The maturing opaque eggs which are large laden with yolk.
3. The atretic eggs which are mature are never shed but are reabsorbed for hormone function.

TYPES OF EGGS

1. **Planktonic eggs:** These eggs float passively eg. *Mackerel sardine* etc.

2. **Semi buoyant eggs:** These neither float nor settle on bottom eg. (eggs of shad).
3. **Demersal eggs:** Eggs sink to the bottom eg. Bitterling, Rhodes.

Condition of Eggs

1. Healthy and fertilized eggs attain the average diameter for the species that look transparent, turgid and hard.
2. Dead eggs or unfertilized eggs appear opaque, small in diameter and irregular in shape.

Number of Eggs

In many fishes the small eggs remain in the ovary after spawning and are gradually reabsorbed. The number of eggs reabsorbed after spawning varies with the age. For example, in carps, the number of reabsorbed eggs is greater in the youngsters in comparison with middle aged and older ones.

Length of the fish in cm.	35	40	45	50	55
% of Small eggs.	24.2	21.5	16.1	15.8	19.9

The fecundity is a quality of the species which ensures high rate of intraspecific competition and more rapid rate of evolution *i.e.* the species with higher fecundity have a greater biotic potentiality.

The concepts of biotic potential and environmental antagonism introduced by Chapman do not reflect any actual biological principles. The fecundity and its adaptive change ensure the preservation of the species and not its extermination. The species responds to the change in environment by change in its fecundity.

CLASSIFICATION OF FISHES ON THE MANNER OF SPAWNING

According to Kryzhanorsky 1949, the fishes are classified on the manner of spawning.

1. **Lithophila:** Fishes which spawn on hard, stony surfaces (Salmon).
2. **Phytophils:** Fishes which lay eggs among aquatic plants. (Carp).
3. **Psammophilus:** Fishes which deposit eggs on sand (Loach).
4. **Ostracophils:** Fishes which deposit eggs inside a bivalve. (Chilo Gobio).
5. **Pelagophils:** Fishes which spawn freely in column of water and the eggs float. (Cod).

In the Manner of Reproduction, Fishes are grouped as follows

a) **Oviparous:** Fishes which lay eggs that hatch externally, fertilization of eggs may take place internally or externally.
b) **Ovoviviparous:** Fishes in which eggs hatch internally but the

young does not receive nourishment from the mother.

c) **Viviparous:** Fishes in which eggs hatch internally and the youngers receive nourishment from the mother before extrusion.

GROUPS OF FECUNDITY

The fishes differing in the value of index and the adoption capacity to various rates of consumption by predators.

- ***I Group:*** (Having a fecundity from 1 to 10). Some fishes such as sturgeon, coregonus etc. which adapt to a relatively low rate of predation.
- ***II Group:*** (Having fecundity from 10 to 50) these fishes like pipe and its allied forms having a competitive predation.
- ***III Group:*** (Having 50 to 100 fecundity) these fishes are adapted to considerable mortality *e.g.* white fish. In each species the individual and specific fecundity is subjected to a great variation, primarily related to the variation in the size of the fishes.

In majority of fishes, the capacity of egg laying gradually increases with the age of fish, but it decreases as an individual approaches sterility. In all species there is an increase in the number of eggs laid with the length of the fish to a certain period.

TYPES OF FECUNDITY

1. Individual Fecundity

1. The fish is measured and weighed.
2. The ovaries are taken out and weighted.
3. The sample of 10 of 15 gms. is taken out from the ovary.
4. The eggs in the samples are counted under a low powered microscope.
5. Number of eggs are multiplied by the weight of ovary.
6. In case of different size of ovaries, average count is taken. For example Sunfish (Molamola) produces large number of eggs. A 54 pound fish of this kind is said to produce 28,000.00 or more eggs in single season. The cod (Gauds) produces as many eggs as 9,000,000 per season. Fulton estimated that a ling contains from 14,000,000 to 60,000,000 eggs. The eggs which are looked after are usually few hundreds. The viviparous fishes still fewer. But Blanng keeps 300 young's in its ovary at a time.

2. Relative Fecundity

The relative fecundity of a fish denotes the number of eggs per unit weight or length of the fish. But for a population, individual and relative fecundity are not always characteristics. The size of population of a species depends on the individual fecundity, the time of sexual maturity, and the periodicity and frequency of spawning of the life cycle of the individual.

According to Severtor (1941), s formula for specific fecundity of a fish is as follows:

$$\text{Fecundity} = \frac{1}{(1 \quad r)} \text{Pjs.}$$

Where,
r = Number of eggs laid in one spawn
p = Time period between two successive spawns.
j = age of fish at maturity.
s = Sex ratio.

Joganson (1955) modified this formula as follow:

$$\text{Pj}\sqrt{\text{rx}}.$$

Where x is the number showing the life of fish. According to Joganson, majority of fishes has the sex ratio of unity. Both these formula indicate reproductive capacity of the species. The capacity of egg lying increases gradually with the age but it decreases as the animal becomes sterile.

ANALYSIS OF FECUNDITY AND REPRODUCTION

Fecundity expresses the capacity of fish in terms of egg production per year. Absolute fecundity is defined as the number of eggs in the ovaries of an individual that are ripening for the coming spewing period. This factor is useful for racial studies in taxonomy and for studies on population and production. It is expressed as the number of eggs per unit length or weight of fish. (Relative fecundity).

Absolute fecundity. F = G/g.
where,
f = fecundity.
n = number of eggs.
G = weight of ovary
g = weight of sub sample.

METHODS OF ESTIMATION OF FECUNDITY

A. Counting the Number of Eggs in Complete Ovary

For counting the number of eggs the complete ovary is taken out and preserved in 5% formalin. By the process of water displacement the volume of ovary is calculated.

The ovary is cut into 3 or more segments. The volume of each segment is determined. By subtracting segment, the volume of other eggs and ovarian tissues from the volume of ovarian segment, the volume of mature egg is calculated. The mature eggs are then counted. This procedure is repeated with each segment. Then average count is calculated. Thus the number of mature egg per c.c. and total number of mature eggs in the entire ovary is found.

Now a days formalin is replaced by modified Gilson's fluid. In this fluid the ovaries are left for 24 hours or several months. Small ovaries are turned inside out while larger cut in small pieces. After the penetration of fluid it is washed properly and ovary shaken well so that eggs are separated and counted. The Gilson's fluid formula is given below

Mercuric Chloride.. 20 gm.
80% Nitric Acid.. 15 ml.
Glacial Acetic Acid.. 18 ml.
Alcohol 60%.. 100 ml.
Water.. 880 ml.

(B) Counting of Laid Eggs

a) Volumetric Method

A sample of 100 eggs from the unknown lot of eggs are taken and blotted to remove excess water. Then volume is determined by water displacement technique. The same is again dried and the total volume is determined. The total number of eggs are calculated by the formula.

$$x = \frac{V}{v} \times n$$

Where,
x = Total number of eggs.
V = Total volume of sample (dried)
v = Total volume of sample (Wet)
n = 100 eggs of the sample.

b) Von Bayer Method

Firstly average diameter of egg is obtained using graduated trough then the number of eggs per quart or per c.c. of the lot of read directly from the Von Bayer's table (1910) and the total number of eggs for the entire lot is determined, after finding out the total volume of the lot.

Table 13.1: Number of eggs, as known diameter per given volume based on Von Bayer's, Table.

Inch.	No. per Quart. (964.4 CC)	No. of per litre	No. per 100CC
0.300	2506	2646	264
0.250	4331	4573	457
0.203	8089	8541	854
0.156	17825	18823	1882

Inch.	No. per Quart. (964.4 CC)	No. of per litre	No. per 100CC
0.140	24661	26042	2604
0.125	34647	36587	3658
0.109	52254	55180	5518
0.093	84130	88841	8884
0.078	142600	150585	15058
0.062	283936	298836	29883
0.046	695223	734155	73415
0.031	2271500	2398704	239870

Fluctuations in Fecundity

The fecundity in a single population may undergo considerable fluctuations in relation to the supply of nourishment. A thin population having a sufficient food supply usually possesses a larger fecundity, for example, Humback Salmon shows a poor fecundity in comparison to other Salmons of the same size where nourishment is better. In the Northern Caspian Sea, the fecundity of the roach of a particular size increase in good feeding years. The fecundity will be increased when the intensive fishing is under taken. The population is reduced and the improvement in food supply is observed. The fecundity differences can be observed in fishes of the same length if kept under different conditions. The substantial differences are observed in the fecundity between closely related species and even in the sub species.

FACTORS INFLUENCING FECUNDITY

1. Feeding

Fecundity may be increased due to the reduction of the passive feeding period which may be due to the decrease in the reserve of yolk in the egg.

2. Density of Yolk

Increase in the density of yolk is directly proportional to the increase in the density of fecundity.

3. Size of Gonads

Increase in volume of gonad also increases fecundity. As for example the fecundity of Amur sturgeon, Huso dahuricus is greater than that of white sturgeons.

4. Ripening of Ovary

Individual fecundity is also increased when the eggs present in the ovary ripen together at any time.

5. Biological Factors

Biological factors include stocking density, length, age, size at first maturity, gonadosomatic index, ponderal index, relative condition factor, ova size and segregation to develop sexual urge for better fecund.

6. Environmental Factors

Environmental factors are light, temperature, running water, pH and physico-chemical qualities of soil, water interaction which could affect the food intake and keep the fish under stress are found to be less fecund.

7. Nutritional Requirement

Nutritional requirement and nutritive qualities or diet also effect fish fecundity for example, when Chinese grass carp eat food along with rice sprouts, wheat sprouts, soybean and peanut dryed, then produce better fecund 57500 eggs/kg body weight compared to the fish fed with corn and rice bran had 38214 eggs/kg body weight.

8. Diseased Fishes

There is less fecundity in diseased fishes but healthy ones produce large number of eggs.

9. Physiological Disbalance

Physiological disbalance due to extrinsic or intrinsic factor affect greatly to fecundities. Jhingran and Sehgal (1978) are of the opinion that fecundity in cold water species are related to the green egg diameter. The fecundity is low with increase is diameter of egg. The number of green eggs per kg. body weight ranging from 1471-1542 when fed with nutritive diet as against 1234-1342 in control.

The parasitic infection and physiological disbalance due to extrinsic or intrinsic factor affect greatly to fecundity as the progress of maturation is also severely affected due to above reason.

IMPORTANCE

Fish fecundity is associated with studies of natural mortality, racial characters and total population estimation and assumed economic importance with certain species.

Application of fecundity is concerned with productivity studies. (Allen 1951)

Some workers have calculated total number of pelagic, plaice eggs and divided this by the average fecundity to give the number of female parents.

Farran 1938 described the differences in fecundities of herring between autumn and spring spawners are probably due to racial differences.

Lyamin (1956) showed that in Icelandic Summer and winter, spawners were morphologically and physiologically similar but their fecundities differed markedly.

Kandler and Pirwitz (1957) found that the fecundity rate in flounder dab, turbot and Plaice were higher in the Baltic Sea than in the North Sea.

Absolute fecundity is called as the total ripe eggs in the ovary of an individual fish. It is related to length, weight or age of the fish and increase with size and age of the fish. Bagehal (1963) observed that in most fishes the number of eggs does not change significantly as the season progresses but the gonad weight increases due to an increase in water content, organic matter content or organic matter transferred from somatic tissue. Hence, the use of regression coefficient in relative fecundity assumes to be in LogF=loga+logb w. which does not differ significantly from unity.

a) Fecundity and Age

The fecundity differences due to variations of age necessitates the importance of age specific fecundity for each age group.

The absolute fecundity of all brooding females and the number of eggs laid by the population in one season is important.

Garking (1967) concluded that variations in individual fecundity is so great that it hides any effects of age it may have. Ludwig and Lange (1975) used age and age length interrelation in his statistical model for estimation of predictive fecundity than length alone.

b) Fecundity and Length

A close relationship is usually found between fecundity and length.

c) Fecundity and Weight

The fecundity is closely related with weight than length. In many fishes the somatic weight changes significantly towards spawning due to nutrient flow from somatic tissue to ovarian tissue. Therefore, the relation between fecundity and weight will differ as breeding season approaches.

d) Fecundity and Egg Size

Fecundity and egg size are correlated. It is generally assumed that fish that lay many eggs must lay small ones Rounsefell (1957) found that the salmonids with the longest fresh water lives have the largest egg such as Oncorhynchus and smaller egg in salmon genus. The genus *O.*

nerka lay smallest and *O. tshawytscha*, the largest. The size frequency of ovarian egg gives the idea of the spawning cycle of a fish. Macer (1974) reported in his observation on some horse mackerel (multiple spawner) that there could be a substantial difference between fecundity and fertility because in such case of multiple spawner, the ovary goes through part spent to recovering state with many eggs unshed. However, nearly all spent fish have some residual eggs. After spawning all spent fish have some residual eggs which is not a significant number compared with the total shed.

e) Fecundity and Spawning Time

The spawning sequence of herring in which fish travel to the spawning ground with the sequence of gonad development. The spawners first spawn are less fecund due to transfer of somatic reserve to gonadal maturity. In carps, the same brooder can be bred more than once by intensive compensatory feeding under controlled conditions. But in high fecundity group of fish, the maturation taken longer period as in carps and in less fecund group of fish as in Tilapia, maturation takes short period.

f) Fecundity and Environment

Any change in the environment may result in significant changes in fecundity and so alter production estimates. An increase in food supply with increase in food intake not only increase the growth rate but also increase the gonad weight due to exogenous flow of yolk for a given age. Therefore, the fecundity for a given length increases.

g) Fecundity and Egg Quality

The eggs of most species vary in size and chemical composition. These variations will be important from the view point of fish production. It is a meaningful aspect of fish production studies.

FECUNDITY OF CULTIVABLE CARPS

There are considerable variations in fecundity in regard to each species, age of fish and ecological condition of habitat. In several species, differences in fecundity is dependent with size of ova, length of fish, availability of food and other factors suggested earlier.

1. *Catla catla*

The fecundity range was found to be 20-246 numbers per gm body weight of pond reared fish and the number of eggs vary from 604 to 1330 per gm. of ovary weight.

2. *Labeo rohita*

The absolute fecundity of several specimen of rohu ranging from 51-57 cm. is to be variable. The number of eggs per gm. of body was 271. The fecundity range was found to be 109-413 numbers per gm. body wt. of fish and the number of eggs vary from 747 to 1528 per gm. of ovary weight.

3. *Labeo calbasu*

The data available for calbasu shows that a female weighing 500 gm. had 2, 19, 450 era. The average being 438 eggs/gm. body weight and 1463 per gm. ovary weight.

4. *Labeo bata*

The fish get mature in nine to ten months. Alikunhi 1956 reported that a specimen weighing 1050 gm. had total eggs of 3,00,000 giving an average of 286 eggs/gm. body weight.

5. *Cirrhinus mrigal*

Both the sexes mature at the age of two year. In a ripe female of 1.47 kg. a total number of 2, 16, 800 eggs were reported giving an average of 147 eggs. /gm body wt. The fecundity ranges from 1.44-1.52 lakhs per kg. body wt. of female.

6. Silver Carp

Alikunhi *et al.* (1963) reported the fecundity of silver carp ranging in weight from 3.0-8.5 kg was found to vary 1,45,000 to 30,44,000 giving an average of 171 eggs per gm. of body weight.

7. Grass Carp

In a specimen weighting 4-7 kg. Alikunhi *et al.* (1963) found total number of 3,08,800-6,18,000 eggs. Sukumaran 1969 reported that on an average the fecundity was 82 eggs/gm. body weight and 610 eggs/gm of ovary weight.

8. *Cyprinus carpio*

The absolute fecundity is reported to vary from 39,400-16,59,000 eggs. The absolute and relative fecundity of common carp in India varies with length group being reported by Alikunhi 1966. According to him a fresh length range of 15-25 cm. had absolute fecundity from 6,360-25,942 and relative fecundity per gm. of ovary wt. range from 517-1,388.

9. Big head

Chinese have reported that the average relative number of eggs per gram body weight is 124 in pond reared species as against the fecundity of 96 eggs/gm. body weight in natural riverine condition.

10. Black carp

It is cultivated in China and the fecundity is 114 per gm. body weight in pond reared species as against the fecundity of 93 eggs/gm. body wt. in natural condition.

11. Mud carp

The average relative fecundity per gram body wt. is 240 in pond reared specimen. In Singhi, fecundity varies between 1100-37,000 in species of 120 to 285 mm in length weighing 11 to 112 gms. The number of hatching produced by *Channa striatus* through artificial propagation ranges from 1304-23180 in *C. punctatus* range from 3560-13,960 and in *C. gachua* ranges from 3059-8089. In *Anabas testudineus* ranging from 91 to 140 mm.

CONCLUSION

It can be said that greater the fecundity the more acute will be the "intra-specific competition. It will lead to the extermination of the old forms and new species will emerge. An increase in fecundity may lead to its own extermination. This notion is not correct. In fact an increase in the fecundity of an individual within the population represents an adaptive response of the population to environmental changes. It does not include pathological conditions. Thus it can be concluded that an increase in fecundity ensures the preservation and not the examination of the species. It ensures its relative stability both in space and time in the event of fairly wide fluctuations in the environmental conditions.

Chapter 14

Composite Fish Culture

INTRODUCTION

A fish pond is a complex ecosystem as the surface is occupied by the floating organisms such as phyto and zooplankton. The column region has live and dead organic matter sunk from the surface and the bottom is enriched with detritus or dead organic matter. The marginal areas harbour a variety of aquatic vegetation. The different trophic levels of a pond are utilized for increasing the profitability of fish culture.

The main objective of the composite fish culture is to select and grow compatible species of fish of different feeding habits to exploit all types of food available in the different regions of the fish pond, for maximizing fish production. Fish production by old traditional methods gives low yield, but if several species of different age groups feed upon different kinds of natural food stocked together in proper proportion in the same pond, so that all available ecological niche are exploited and the production increases several times. This is based on the principle that compatible species do not harm each other and there is no competition between different species for natural food. This may have a natural effect on the growth of others.

The common species of carps having compatibility and different feeding habits include the Indian major carps *i.e.* catla, rohu and mrigal and exotic carps such as common carp, silver carp and grass carp.

Recently, the fringe-lipped carp and the milk fish which are common in brackish water have also been incorporated in the composite fish culture system. The stocking density of fingerling is commonly followed, depending on the fertility, environment, level of feeding, and availability of fish fingerlings. The stocking percentage of the different species may be determined for adaptation. (Table 14.1)

Table 14.1: Percentage of different species in composite fish culture

Type of fish	4 Species Combination	5 Species Combination	6 Species Combination	8 Species Combination
Catla	30	30	10	10
Rohu	15	15	15	10
Mrigal	30	25	20	20
Common Carp	25	20	20	15
Silver Carp	-	-	25	25

Type of fish	4 Species Combination	5 Species Combination	6 Species Combination	8 Species Combination
Grass Carp	-	-	10	10
Fringe-lipped Carp	-	10	-	05
Milk fish	-	-	-	05

DEFINATION

Stocking of fishes of different habits, in same pond is called polyculture or composite fish culture.

HISTORY

China was the first to promote composite fish culture

1. The black carp (*Mylopharyngodon piceus*) that lives on snails and other molluscs at the pond bottom.
2. The grass carp (*Ctenopharyngodon idella*) which feeds on coarse vegetable matter.
3. The silver carp (*Hypophthalmichthys molitrix*) which depend son planktonic organisms.
4. The big head (*Aristichthys nobilis*) eats macro plankton.
5. Common carp (*Cyprinus carpio*) which is an omnivorous scavenger.
6. Bream (*Parabramis pekinensis*) which is an omnivore.
7. Mud carp (*Cirrhinus molitorella*) is a bottom feeder.

The Chinese composite fish culture, described by Hora and Pillary (1962) is classified into three groups.

a) The Kiangsu and Chekiang.
b) The west river regions.
c) The Hong King system.

The Kiangsu and Chekiang

In Kiangsu and Chekiang system the stocking evolved in relevance to the nature of local ponds and the availability of stocking material. The black carp is preferred to grass carp, because the great market value and easy availability of pond, snails its choice food. The cold climate area precludes the culture of mud carp.

The West River Regions

In the west river system the stocking methods are modified to suit local conditions. In warmer climate, depth of pond, nature of water supply, kinds of food available and basic pond fertility are the factors for determining the stocking rates and combinations of fishes for example, the Black carp, Silver carp, Big head and common carp are cultured in pond.

The Hong Kong System

In this system, the distinctive feature of pond stocking, the use of grey mullet which are abundant, as the climate in the region is warmer, black and mud carps are less used. Slightly brackish water is used in ponds and since such ponds are heavily stocked, regular cropping is practised as needed to thin the stock. The species combination, stocking rates etc. are practised in China, example the grey mullet Grass carp, Silver carp, big head, mud carp, common carp and bream.

TYPES OF COMPOSITE FISH CULTURE

Indigenous Species

In India composite fish culture is an old practice. *Cirrhina mrigal* and *Labeo calbasu*, are bottom feeders that feeds on decayed vegetation animals, plants and epiphytic plankton from the bottom of the pond, can be cultured with each other. Catla is a surface feeder and feeds on zooplankton. *Labeo rohita* is a column feeder and is generally stocked together in the same pond. According to Alikanhi (1957) Catla, Rohu and Mrigal are stocked in the ratio of 3:3:4 to give good yield from the ponds. But Hora and Pillay 1962, have recommended the following stocking proportion:

Catla	30	30
Rohu	60	50
Mrigal	10	10
Kalbasu	x	10

Stocking of these fishes in Indian Carps in same pond is an excellent example of the correct selection of species for the maximum utilization of the food from the different zones of the pond. Culturing of a single species is not profitable, therefore, a group of fishes with different feeding habits are selected.

Exotic Species

Exotic carps, *Hypophthalmichthys molitrix* (Silver Carp), *Ctenopharyngodon idella* (Grass carp) and the *Cyprinus carpio* (Common carp) have been selected for culturing the major carps. Silver carp is surface feeder that feeds only on phytoplankton, unlike catla which feeds on zooplankton. The common carp is omnivorous bottom feeder while the grass carp (Column feeder) is known for efficient utilization of all noxious and macro vegetation not utilized by any other fish, rather its, faecal matter. Further is serves as food, for the common carp and also accelerates plankton production and growth of the fishes.

The Indigenous and exotic fishes such as *Catla-catla* (catla), *Labeo rohita* (rohu) *Hypophthalmichthys molitrix* (Silver carp) *Ctenopharyngodon idella* (Gross carp), *Cyprinus carpio* (Common carp) is cultured together, then the production increases eight to nine times. Researches by the Central Inland fisheries of Indian and exotic species gave best results. The observation are:

a) When the Indian major carps alone were cultured, the average yield was 2088 kg/hec./year.
b) When exotic carps alone were cultured, the average yield was 2,900 kg/hec./year.
c) A combination of Indian and exotic carps gave an average yield of 3,085 kg/hec./year.

The stocking rate and proportion of fingerlings of different species depends on the composition and abundance of various fish food organisms. The rate of fertilizer and manure also depends on the physico-chemical conditions of water.

COMPOSITE FISH CULTURE

During the recent years composite fish culture has been taken up in large sized ponds as at Anjana fish farm, West Bengal, an average production of 4000 kg/hec./year was obtained. In CIFRI at Cuttack a record yield of more than 9000 kg/hec./year was obtained. The following species were used in the ratio:

Grass Carp 25
Catla catla 15
Silver Carp 15
Rohu 30
Scale Carp 40

Thus, by applying modern methods of fish culture using suitable combinations of fish species mixed culture in the fish production can be achieved in the country.

In Southern India, *Labeo fimbriatus, Labeo kontius, Cirrhinus cirrhosa,* and *Catla calta,* are cultured in varied ratio. In Bihar, West Bengal, U.P., Haryana, Maharashtra, Andhra Pradesh and Tamil Nadu a mixed farming on the basis of ecology has been encouraged. These indigenous and exotic major carps are chosen for research project.

Table 14.2

Indigenous Species	Species ratio	Habit
1. *Labeo rohita.*	25	A column feeder feeding on decaying plants.
2. *Cirrhinus mrigal*	1.0	A bottom feeder consuming decaying plant, detritus mainly.
3. *Catla catla*	1.0	A surface feeder which consumes zooplankton.
Exotic Species:		
4. *Cyprinus carpio*	2.0	An omnivorous habit scavenger on plants and animal.
5. *Ctenopharyngodon idella*	1.0	Feeding on coarse macro vegetation.
6. *Hypophthalmichthys molitrix.*	2.5	A surface feeder on phyto-plankton.

The average yield of indigenous major carps in mixed farming has been estimated to be 4000k/ kg/hec./year and the exotic carps is 3000 kg/hec./year.

Stocking density of six species has now been increased to 8000 kg/hec. with the help of intermediary input technology and 10,000 per hectare with the help of high input technology.

Culture practices and many more combinations were formed and proved to be a success. Hora and Pillay (1962) proposed culture of catla, rohu, mrigal and calbasu (L. calbasu) in proportion of 39:50:10:10 composite fish culture at Cuttack involved a maximum of 7 species viz. the grass carp, silver carp, catla, rohu, scale carp, mirror carp and the exotic genus Tilapia in proportion of 25:15:15:30:40:10:15 respectively. Many more experiments conducted by Central Inland Council of Agricultural Research (ICAR) have high yield and income. Various combination selected for experiment chiefly include.

1. Grass carp, Silver carp, scale carp and mirror carp in ratio of 5:3:8:2.
2. Grass carp, Silver carp, rohu, scale carp and mirror carp in ratio of 5:3:6:8:2.
3. Grass carp, Silver carp, catla, rohu, scale carp and mirror carp in ratio of 5:3:3:6:8:2.
4. Grass carp, Silver carp, catla, rohu, scale carp and mirror carp and tilapia in ratio of 5:3:3:6:8:2:3.

Feeding

The grass carp normally feeds on aquatic weeds (Hydrilla, Najas, Ceratophyllum and chara etc) forage grass (Napier grass) topical leaves and kitchen vegetable.

They are fed twice a day at the rate of 100 kg/hec. in the first month and the quantum is increased by 100kg/month at fortnightly or monthly intervals, till the end of harvesting. The food of grass carp is normally placed on floating frame made of bamboo poles. The powdered groundnut oilcake and rice-bran may be mixed in the ratio of 1:1 and soaked in water for some time. Then it is made into small balls for the purpose of feeding. Further, groundnut oilcake, rice-bran, sieved flour mill waste, kitchen waste and chopped and boiled vegetable waste may also be used as fish food. The food balls are placed in bamboo baskets in shallow areas (0.5 m depth) of the pond in three or four different places.

To enhance the growth of fish and minimize the period of rearing, protein-rich feed components such as silkworm pupae, soybean, cowpea, notonectids (aquatic insects), trash fish a prawn waste may also be incorporated in the feeds. It is reported that artificial fish feed containing 40 percent protein is suitable for a good growth of carps.

The supplementary feed is given initially at the rate of 1% of the total body weight of culturable fish and increased 2-3% depending on the utilisation or based on the total weight of fish. This artificial feeding is to be suspended along with manuring, if and when, the green water associated with algal blooms is noticed.

RECENT RESEARCH IN COMPOSITE FISH CULTURE

In India the composite fish culture has been taken in large sized ponds at the Anjana fish farm, West Bengal an average production of 4000 kg/hec./year has been observed. At Gujarat fish farm (Jaunpur U.P.) the average yield 3000 kg/hec./year has been obtained. In the CIFRI at Cuttack record yield of 9000 kg/hec./year was obtained. Some experiments have also been conducted on integration of aquaculture with livestock rearing to reduce the cost of fish production.

In a fish cum-duck rearing centre a yield of 4323 kg/hec./year was obtained without adding any fertilizer and without giving any supplementary feed to the fish. In fish cum pig culture, yield was 7300 kg/hec., was recycled through the pond. In mixed culture the maximum production of M. Malcoemsoni was 30,000 kg/hec. and the stocking rate was 337 kg of prawn and 2084 kg of fish.

The Jute Agriculture Research Institute Barrackpore, where the chloride content of pond water ranged from 36 to 88.5 ppm a production ranging from 2785 kg/hec./13 months was demonstrated in three ponds using the Indian major carps. (Fig 14.1).

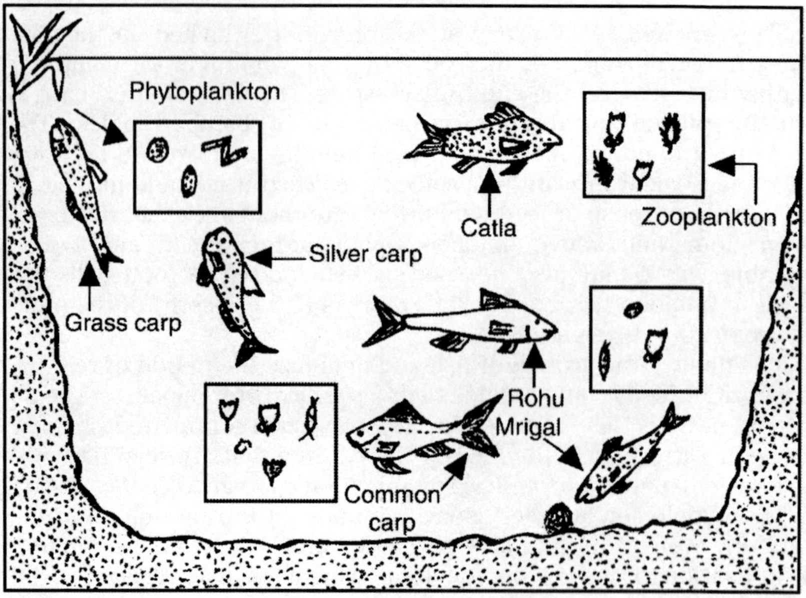

Fig. 14.1: Trophic tiers in composite fish culture.

RECENT RESEARCH CENTRES OF COMPOSITE FISH CULTURE

All India Coordinated Research Project on composite fish culture of Indian and exotic fish was initiated in 1971 by the Indian Council of Agricultural Research, with its headquarters at the Central Fisheries Research Institute. There are six centres in different states:

(1) Andhra Pradesh (2) Haryana (3) Maharashtra (4) Tamil Nadu (5) Uttar Pradesh (6) West Bengal.

The main results of composite culture of Indian and exotic species at two centres:

a) 2 ponds at Gujarat fish farm (Uttar Pradesh) gross and net yields to 2,692-3,210 kg/hec./6 months and 2569-3087 kg/hec./6 months respectively.
b) At Kulia fish farm, Kalyani (West Bengal) gross and net yields of fish harvested from one pond confirmed to 3,232 kg and 2,917 kg/hec./6 months.

The research project on composite fish culture and fish seed production have indicated possibilities of further increasing fish production at all the centres.

i) At the Karnal Centre (Haryana) production as high as 3448 to 5894 kg/hec./6 months were obtained with the six species composition.
ii) At the Poona Centre (Maharashtra) production with six species

combination increased from 2199-2242 kg/hec./8 months to 10,148/ kg/hec./year.

iii) At Bhavani Sagar Centre (Tamil Nadu) gross production around 3000 kg/hec./year. This production was less due to minnows and murrels which always gained an easy entry.

iv) Kalyani centre (West Bengal) the production ranged between 3397 and 6053 kg/hec./year.

v) Badampudi (Andhra Pradesh), the production between 3535 and 4648 kg/hec./year.

vi) Ranchi centre (Bihar) the production has increased from 1504 kg/hec./hear to 3526 kg/hec./year.

vii) Gauhati centres (Assam) a high rate of production of the order of 4084 kg/hec./year was obtained.

viii) Kaushalyaganj Centre (Orissa) a production of the order of 4107 kg/hec./6 months was obtained.

TECHNIQUES OF COMPOSITE CULTURE

In recent years, the two sizes of common – carp are co-stocked and in the other either tilapia or grey mullet or both are cultured with common carp and in other experiment the culture of silver carp with other fishes have also been conducted. Fish pond in Israel, during the growing season are systematically fertilized at bi weekly intervals with ammonium sulphate or liquid ammonia, super phosphate and poultry manure. The fishes are given artificial feed daily. Of all the fish species tried in mixed culture with carp Tilapia aurea and Mugil cephalus have proved to be the best. Tilapia aurea has been observed to increase fish production by 30%. Yashouv (1967) suggested a method for evaluating the efficiency of mixed fish culture over that of monoculture in terms of increased fish production in ponds. In the case of mixed culture an index of competition, as suggested by him can be estimated as follows:

$$C = (a-b)/a.$$

C = Index of competition.

a = Daily carp production in mono culture.

b = Daily production in mixed culture.

In pond culture, the interrelationship in fish populations is satisfactory. Crops of harvestable fishes that are satisfactory in amount when the basic fertility of water bodies containing the populations is taken into account. Such populations are considered to be balanced populations. The characteristics of a balanced populations are as follows:

i) A define range in the ratio of the weights of forage and piscivorus groups.

ii) A narrow range in the ratios of weight of small forage fishes to the weights of piscivorus groups.

iii) More than 33% of total population weight in the form of fishes of "harvestable size".

Swingle (1950) has described the techniques for determining the interrelationship and dynamics of balanced and un-balanced fish population. He described six criteria.

i) *F/C Ratio*: The population may be divided into two classes.
   a) The 'C' Class composed of species that feed principally upon other fishes and that cannot attain normal adult size without such food.
   b) The 'F' class composed of all other species that feed upon plants, plankton, water insects and other small aquatic invertebrates. The 'C' value is the weight of the 'C' class and 'F' value is the weight of 'F' class and F/C is the ratio between the total weights of forage fish eg. blue gills and the total weight of the all piscivorus fishes *e.g.* bass in a population. In balanced populations the most efficient were in the range where F/C quelled 3 to 6; F/C values range 1.4 to 2.0 were obtained in crowded population.

ii) *Y/C Ratio*: The Y/C ratio is the ratio of the total weight of small fishes in the 'F' group and total weight of 'C' species. The Y/C ratio is an expression of the food available to the 'C' class. In balanced populations Y/C ratio ranged from 0.02 to 4.8. All population with Y/C = 0.2 to 0.5 were crowded with 'C' species.

iii) *AT Value*: (At value means availability value) The AT value is the percentage of total weight of a fish population, composed of fish of harvestable size. The most remarkable populations had values 60 to 85 and values above 85 indicated the overcrowding of population with 'C' species. Unbalanced population when AT value range 0.0-4.0.

iv) *AF Value*: AF Value is the percentage of the total weight of the 'F' classes composed of large fishes. The IF and SF values are the percentage of the total weight of 'F' Classes composed of the intermediate and small fishes respectively. The most desirable population were in the range when A/F Values ranged from 60 to 80. The unbalanced population had AF values from 1.2 to 25.7.

v) *IF Value*: The IF values overlapped in balanced 41.4 and unbalanced populations 0.04 to 100.4.

vi) *SF Value*: The SF value in balanced populations ranged from 0.4 to 80.9 satisfactory population occurred when SF values ranged from 15 to 40. Values in excess of 60 indicated inefficient populations.

The AF, IF and SF values were found to be shifting with changes due to harvest, predation, natural mortality and growth.

Factors

The factors which are considered useful for the purpose of further utilization of natural, and supplementary food and higher productivity are given below.

The selection of suitable fishes of carp family, is generally done. It is important to note that all these fishes generally feed on a similar type of food. Their capacity for reaching out for food as well as the feeding inches are not identical. Hence, the adjustment with respect to the fishes selected for the purpose.

1. Selection of suitable combinations with respect to food and feeding habits and their tolerance limit to various abiotic and biotic factors.
2. Stocking densities of fingerlings stocked per hectare area of the pond.
3. Initial total weight of fingerlings being released in the pond.
4. Rate, kind and frequency of fertilization of pond.
5. Rate, kind and frequency of artificial food being supplied.
6. The physico-chemical and biological characteristics of the pond.
7. Fish production including the gross weight of fish harvested per hectare area of the pond.
8. The cost of production, based on expenditures incurred the net profit.

Economics of composite fish culture of fish ponds vary in their water, soil qualities and method of culture operation. It may not be possible to generalize the nature of fish production and its cost functions. However, the economics worked out on the basis of experimental composite fish culture in a 0.1 hec. pond is given below.

Economics of Composite Fish Culture

Items	Amount (0.15 hec).
Operational Cost.	
a) Cost of fingerlings (1,000)	Rs. 200
b) Cost of chemical fertilizers and organic manures.	Rs. 120
c) Cost of supplementary feed.	Rs. 625
d) Miscellaneous expenses.	Rs. 100
e) Interest on capital invested at 10% (Value of pond, Rs. 4000)	Rs. 400
Total	**Rs. 1445 or 1450**
Gross Income:	
Sale proceeds of fish harvested (375 kg) at Rs. 8/- kg.	Rs. 3000
Gross income	Rs. 3000
Operational Cost.	Rs. 1450

Items	Amount (0.15 hec).
Net Income/0.1/hec/annum.	**Rs. 1550**
Estimated income/hec. Rs. 15,500/-	

CONCLUSION

Different types of fishes are cultured in a pond. Bottom, column and surface feeder fishes feed on different regions of the pond for the maximum yield. There is no competition between them for different types of natural food, thus their growth increases giving maximum benefit.

Chapter 15

Fish Marketing

INTRODUCTION

The term fish marketing was used to denote simply buying and selling of fish at a landing centre or the beach side. Recently, the fishermen used to sell fish according to whatever demand that was existing either in the quay or nearby place. The present system of marketing not only thinks of meeting the effective demand for products but also creating demand in the market. Fisheries have become highly industrialized in all advanced fishing nations. The new marketing techniques have been adopted so as to sell more fish not only in local areas but also in distant and interior markets.

Modern fish marketing system in meeting the existing demand for fish, besides tapping the potential demand in the important markets. Changing the fish form according to the taste and needs of the consumers would result in mere sales, thus the market has to design the new product features developing, packing and brand name pricing the product to get a better result on investment. Constant advertising has to be done for the awareness about new products so as to gain highest amount of output.

Fish market is a place inside or outside the country where the fishes and fish products which are of commercial importance are subjected to sale and it is a link between the trader and the consumer. A good market always serves to safe guard the interest of the trader and the consumer. A trader also cares for maximum production, best possible quality of commodity and their timely, supply to consumers at, reasonable cost. Fishes are directly associated with the economy of the country, it earn foreign exchange also by the disposal of the commodities in the international market. The adoption of the modern technological methods in the fish marketing managements which means more sales of fish and reducing further the wastage of fish, more fish there by will be available for human consumption. In short fish marketing system based on modern lines has to be developed in tall developing countries, including India. Fish markets of many places are controlled by middlemen or powerful rings of businessmen with the result that the fishermen get very low price for their product even when the consumers pay highly for it.

DEFINATION

Fish market is a place inside or outside the country where the fishes and fish products of commercial importance are subjected to sale. These fish production and consumption is properly regulated, is known as fish marketing. The marketing is integrated part of fish industry.

Characteristics of Fish Marketing

1. Perishability of the commodity.
2. Seasonal concentration of landing (glut).
3. Scattered landing places (great distance of transportation).
4. Small quantity produced by each fisherman is to be assembled, sorted and transported.
5. Non-uniformity of landings, quantity, freshness and size.
6. Handling and treatment, icing, freezing, curing, packing.
7. Large number of intermediaries involved.

AIM OF MARKETING

From the nutritional point of view only fresh fish should be provided. This work engage more labors. Each agency must be happy in the interest of producer and consumers.

Present Status of Marketing

Present status indicates limited transport facilities. The distribution of fishes is greatly distractor from land centre. Fish marketing should be fairly localized and loosely linked in different areas. Relatively for human beings, fishes are short product. Physical distribution, adverse geographical conditions and climate conditions fluctuate the qualities of fishes. Middlemen provide help to fishermen in transportation and given financial add also but because of financial riser, retailers and wholesalers get more advantages, so that as there is less production and supply of fishery and fishes, therefore fisherman gets low price and, buyer get fish in "high price".

Important Points Related to Success of Trade

1. The demand position of the consumer market should always be assessed.
2. Sudden price rise and also fall of price should be prevented. Prices should be stabilized irrespective of bumper or poor landing of fish and fish products. Creating of artificial scarcity leads of dangerous consequences.
3. There is great need for quality control at various levels as per requirements of the local and foreign markets.

4. All precautions should be taken to prevent fish and fish products due to wastage.
5. A preliminary statistics on catches should be collected prior to diversion of the fish to various agencies dealing with storage, preservation, canning industries, and manufacturing units of various products and by products. Statistics of fresh consumption should also be collected.
6. There is need to be alternative to even small points of conscience against the harmful activities and functioning of the middle men, money lenders, hoarders retailers and even co-operative societies are involved in this trade.
7. That the fishes have very high food value being rich in protein, carbohydrates, vitamins and minerals, should be given wide publicity.

TYPES OF MARKETING

Traditional Fish Marketing

The traditional system of marketing is generally followed in which all methods and practices in trade dealing are based on some customer that remains unchanged and unimproved over decades. Normally a woman who squats on the floor with the fish around her. She hides the good quality of fish and tries to sell the fish in the first instance that had already spoiled. The fish is usually sold as a whole piece as traditional practice. There is no free entry in the fish marketing since old fish merchants do not allow new persons into the trade. It is usually these markets or producing centres are faced either with monopsonistic or monopolistic methods of control over fish supplied.

Modern Fish Marketing

Fresh fish cannot be kept for a long time, it has to be preserved by proper means for enhancing the marketability of product. Sometimes the fish in its original form after its capture may not be immediately demanded by consumers but when processed into other forms would find already market, for example ocean perch fish in the U.S.A. did not find a ready market but when it was frozen there was a good demand for it. The in marine products processing centre, college of fisheries Mangalore, fish sausages which were processed with fresh fishes like sciaenid's and sharks, were much demanded even from the classes who never consumed fish before likewise, CIFE, Mumbai had developed technology to manufacture fish wafers, similar to potato wafers.

CHANNELS AND STAGES OF FISH MARKETING

1. Purchase of fish at landing places by fish dealer or middlemen either from fisherman or vessel owner often fisherman are in debt and obliged to deliver.
2. Assembly of small quantities of fish into large shipments sorting bulk breaking and transport to wholesalers.
3. Consignment sale by whole salaries normally by means of auction. There are several methods of auction, such as ascending, descending, whispering, tender commission 5-10% of proceeds charged.
4. Retailing in fish shops, fish managers, super markets.

METHODS OF SELLING OF FISH

1. In the primary fish markets, the fisher folk directly sell fish to the buyers without having any intermediary between fishermen and buyers. In many places the buyers may be wholesalers, dry fish merchants or middlemen. In Tamil Nadu, Andhra Pradesh and Maharashtra the women who belong to fishermen's families, traditionally sell fish in retail markets unlike the upper eastern or northern India. The prices are usually negotiated between individual fisherman seller and buyer.
2. The fishermen send fish to the commission agents as consignments. The commission agents or "aratdars" auction the fish and the gross sale proceeds are remitted to the fishermen after deducting the various marketing charges. This system of selling is known as auction system. In this system the fishermen is obviously at a disadvantage since he will not be available at the time of auction. A few cooperative societies have come up in the place of agents for marketing the catches in the fish markets so as to strength the bargaining position of the fishermen.
3. Selling is also done by contract method. The prices fixed before the fishing season starts. The traders make some agreement for the delivery of catches by fishermen at stipulated prices. All the fish caught is disposed of at the fixed price.

   In Gujarat, the "Mandala" committees is formed where the fishermen's representatives, government representatives and traders assemble and fix a price based on the previous year's prices and also to some extent on the Mumbai fish market prices. In this case prices do not generally fluctuate very much as any amount of quantity is purchased by the traders at the price agreed by the "Mandala" committee.

Normal Cost Structure

Roughly speaking, producers, dealers other than retailers and retailers share respectively about 45%, 23% and 32% of the price paid by consumers.

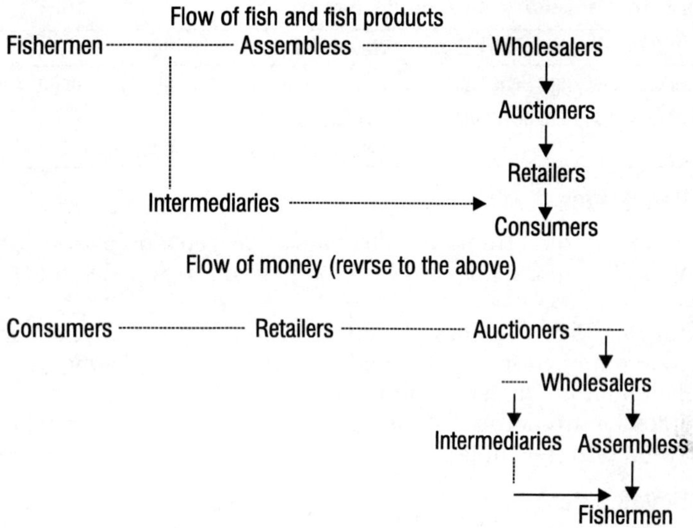

Fig 15.1: Flow of fish and fish products

Each pays necessary expenses and obtain price margin or commission.

Factors affecting Fish Marketing Patterns

a) Concentration of landings as a result of construction of improved part and landing facilities, increased maneuver ability of boats in the course of fishery development.

b) Increased size of capital requirements for fishing activities often with outside investors coming to fisheries. Increased scale of operation resulting from economic growth. The above factors will lead to greater bargaining power of fishermen or producers.

c) Increase of processed or frozen fish.

d) Institutional development by government intervention *e.g.* fish marketing organization, state fishing corporation, government loan scheme and fishermen's cooperatives, very likely to affect and change fish marketing patterns.

Core Fish Marketing System Analysis

The core fish marketing system is classified the three types:

INTENSIVE GROWTH

Intensive growth is possible when the firm seeks to expand its efforts which are within the premises of the firm. Some opportunities, which are not latent in its present marketing efforts have to be explored, it can adopt product market expansion matrix.

	Present Marketing	New Products Marketing
Present markets	Market Penetration	Product development fish filters, fish fingers.
New Market	Market development	Diversification of sales. Fish wafers and fish sausages.

Market Penetration

a) It consists of attempting to have increased sales for its commodities in those markets through more aggressive marketing effort.

   Example: Price incentives and explaining nutritional values of fish communication may be made by way of advertisements to the public about the high protein value of fish.

b) Samples of prepared fish food in most palatable form can be distributed to the public.

Market Development

Efforts of present fish sales from those markets can be diverted to new market, where new sales centres can be opened in different areas. According to consumer tastes, it is necessary so that new segmented markets maybe approached for further marketing of fish products.

Products Development

It consists of increased sales by developing new products or getting new types of fishes, for those markets which we are attempting improved products like fish bournvita, fish biscuits etc.

Integrative Growth

Integrative growth comes when it takes integration. Some economic advantages from the other firms which have already established their markets in some phases.

The basis unit may combine with another concern by making backward, forward or horizontal integration.

a. Backward Integration

A firm selling fish may seek ownership in other firms who are supplying fish or its products, this is known as backward integration.

b. Horizontal Integration

It may combine hands with its competitors. It may join with some distributing firms who have better control of distribution of fish. Sometimes they may seek the help of middlemen who have a well-organized network for distribution of fish in the country.

c. Forward Integration

It means combining with other firms for distribution of fish for instance, the Amul, a cooperative concern had given its distribution of dairy products to the Voltas. A fishery co-operative producing fish can combine hands with MAFCO a cooperative for distribution of fish.

d) Diversification of Growth

If a core marketing system does not show development and there are opportunities that the laying outside the core marketing, the firm may take up diversified products for marketing. These diversified products, however, should be related to the product marketed.

1. Concentric Diversification

These products that have technological and marketing synergies with the existing fish commodities sales is known as concentric diversification. For instance "fish ham" and "fish sausages" may create an appeal to the customers. These changes will appeal to intermediate customers also.

2. Horizontal Diversification

This is adding new products that will bring some appeal to the present customers eg frozen fish kheema may be supplied along with chilled fish.

3. Conglomerate Diversification

It consists in a firm seeking to add new commodities to new classes of customers because this will offset some of the deficiencies of old products. For example, in India we can substitute some fishes with other fishes whose properties are near or similar to the one that is marketed by the firm when Rohu is not available it can be substituted with common carp or silver carp.

MARKETING ORGANIZATION

Agents and retailers are not suitable for fish exportation, storage and purchase. There is an organisation of fish market.

1. Local assembled market: This market is found near the fish production centres of the coastal villages.
2. Primary market: This is also a small market but here some brokers also take part in it.
3. Central market: This type of markets are in cities that have larger building for storage, preservation and attractive shops.
4. Future market: This market is being planned by the Government.

Such markets shall be regulated by co-operative societies. Suitable facilities for stocking exporting and transporting will be provided by the societies, financial helps shall also be available to poor fisherman to survive. They shall also be trained by the trainees of these societies for better fish production.

Function of Organization

1. Competitive situation of market.
2. Willing of sale product to fishermen.
3. Transport whole selling processing drying to provide salting facilities. This society works on common risk, only trained person works in society, trainers suggest the fishermen.
4. Local co-operative is made for solving money problems. This co-operative provide money to fishermen in low interest.

Marketing Improvement

Improvement in fish marketing system and distribution can very well eliminate some of the depressed pockets of malnutrition, where people live on subsistence level, by supplying fish at reasonable prices. In interior places, where marine fish is not produced it can be marketed after lifting it from surplus coastal landing centres. Fish may not be consistently demanded by few customers due to its peculiar smell. It is the job of a processing and marketing specialist to produce new products made of fish and finally induce them to take fish consumption by explaining its nutritional values. The poor fisherman are being exploited in many Asian and African countries by the middlemen in this trade. These middlemen lend money to the fishermen on very high interest, particularly at the time when there is fall in catches. But now the picture has changed and the credit facilities have been extended to them by co-operative societies at a low interest. Now the fishermen can get education and training at several centres under extension services and rural integration programmes through co-operative societies. The fishermen can be benefited in several ways ex.

1. Facilities for credit, finance and loan.
2. Facilities for distribution, transport and marketing.
3. Facilities for processing and preservation of fishes.
4. Facilities for equipments (Crafts and gears).
5. Facilities for making them aware of the socio economics facts for the welfare of the fishermen and their family members. The socio-economic facts include rights of fishing lease of water, insurance against-risk of life and property, unprecedented poor catches etc. The traditional marketing prevails. It is easy for the producer or any member of his family to approach a consumer and sell the

product directly. Definitely this is advantageous for the producer but producer has no time, as he is mostly concerned with fishing, to market his catches is distant markets. In such circumstances, the distributive machinery has to be tackled by someone else, particularly by any other bigger commercial organisation which can invest bigger capital outlay. There is possibility in improving the economic conditions of fishermen. The involvement of middlemen in the trade is much necessary in the big network of distribution of fish, particularly for a country like India. It may be replaced in the long run when better and progressive machinery and proper institutional frame work are developed.

Co-operative movements have also gained importance in fishery trade. It can safe guard the interest of the small entrepreneurs and stops their exploitation by the middlemen. The fishery trade is also promoted by this movement. There is need for collective efforts by the fishermen, consumers, industrialists and government authorities. In several countries there is much more mechanization and sophistication in fishery operations at all stages right from capture to marketing. Fishery trade can solve many socio-economic problems as poverty hit fishermen. They can be provided many credit facilities for improvement. In many foreign countries – (Canada) loans are granted to the poor fishermen, under "fisheries improvement loan Act".

Suggestions to improve – the mode of marketing

1. By supply a good quality of fish.
2. By maintaining the nutritional value of the fish.
3. By exporting fish to other countries to get foreign currency.
4. By supplying fish to the fish agencies so that consumers may get fresh fish.
5. By organization – Balance between producer and consumer play an important role in the production of fish. To increase fish production, a cooperative organisation being formed by Government which is the association of producer.

Problems

The fish marketing is not an easy task as it has to face many peculiar and special problems at different stages of production and marketing management with greater uncertainties in fish production. The high perishability of fish, assembling of fish from too many scattered coastal landing places, too many species and therefore, too many demand patterns, wide fluctuations in prices adjustment of supplying fish evenly without affecting the demand position quality of fish, last but not least transportation of fish in specialized means of vehicles are the greater sources of irritations and bottles necks, which often make fish marketing much more difficult, the problems included above first hand prices at the auction spots, the final prices paid by the consumers marketing expenses required for the transportation of fish, icing and frozen costs are also important determinants in marketing expenses. It is also necessary. To see what is produced is matched with the demand so that regularity of fish supplies can be maintained as for as possible. It means, not only meeting the present demand for fish but influencing and creating new demand for the product seasonality of fish production though it is sometimes difficult to control also limits the possibilities for marketing of fish throughout at the year unless it is frozen, dried or smoked or canned for instance. The Bombay duck in India, available near the Bombay coast is mostly caught during October to January in a year. The problem of marketing the fish in India is rather complicated and is necessarily to be viewed from an overall angle. Saurashtra and Chilkelake are important fish producing centres.

Marketing Positioning

To, select a specific pattern of market concentration that will give maximum opportunity to achieve its main objective is known as positioning. The firm cannot go to all places even if there are opportunities. It should go after variable positions which should follows the target marketing. It has four characteristics:

1. Market segment is of sufficient current size.
2. It should have potential for growth.
3. Segmented market is not ever occupied by existing competition.
4. Segmented market has some relative less satisfied needs in which a particular firm can serve well.

Important Functions of the Fish Marketing

Fish marketing system plays on important role in the fishery trade. Some important points related to success of this trade are:

1. The demand position of the consumer market should always be assessed.
2. Sudden price rise and present prices should be stabilised irrespective of bumper or poor landing of fish or fish products. Creation of artificial scarcity leads of dangerous consequences.
3. A preliminary statistics on catches should be collected prior to diversion of the fish to various agencies dealing with storage, preservation, canning industries and manufacturing units of products and by products.
4. To determine the proportions of fish landed for diversion to various agencies dealing with preservation, storage, curing, yards, canning industries and manufacturing units of various products

and by products to markets for fresh consumption etc.

5. To establish standards for quality control at various levels of storage, preservation and production of products of by products in accordance with requirements of local and foreign markets.
6. To present losses due to wastage.
7. To keep a strict, vigil on unscrupulous elements and their activities in the trade middlemen. Co-operative societies, money lenders, hoarders and retailers engaged in the trade are to be controlled for their harmful activities or functioning.
8. To educate the people about the importance of food of fish being rich in protein, carbohydrates, vitamin and minerals, should be given wide publicity.
9. To explore the possibilities of export and import there by to help the country to earn more foreign exchange and cooperate in international trade

MARKETING IN INDIA

In India marketing of fish needs too many improvements. Fish markets at many places once controlled by middlemen or businessmen, so that the fisherman get very low price. Co-operative marketing has not made much headway, except in one or two states. Co-operative marketing in Orissa was introduced in 1949-50 in the Chilka lake area. A number of co-operative societies have now formed themselves a union to look after the marketing of fish caught in the Chilka lake. Marketing of fish in Saurashtra is being organized through co-operatives. The object is being to see that the bulk of fish landed in the state is marketed through this agency only.

The Government of Uttar Pradesh is also taking steps to organiser the fish market on a regional basis and to encourage the co-operative organisations to undertake this work. In Mumbai also a union of co-operative societies is managing the marketing of fish by the members of the affiliated societies.

Mumbai is a large producing a consuming centre and exports fish to other states. Kolkata and the adjoining areas of West Bengal are extremely important as large centres of consumption. Catches of fish in many areas along the west coast are large, a small part of this is transported to channel and other large towns for consumption as fresh fish but an account of inadequacy of transport facilities to bulk, find their way to the nearby curing yards to be processed and marketed as dried fish in India and abroad. Taking into these factors and similar other factors, efforts are being made to prepare schemes for the marketing of fish on all India basis, in co-operation with the state Governments and with the help of marketing experts.

CONCLUSION

India is among the ten fish producing countries of the world with regard to sea food production and export. It is keeping pace with other nations in fish production. But the fish markets in different parts of India are not well organized. There is need to reform the markets by modernizing the traditional fish marketing methods by introducing new management techniques. For improving fish marketing the major segments of study should be (a) Demand (b) Supply (c) Attractive marketing (d) Economics. The demand and consumption of fish depends on the availability and also the fish eating population in India. The strategy of fish market management should be analyzed in the light of the present pattern of marketing, setting of objective development of fish demand, formulation of new plan, marketing operation and control etc.

Chapter 16

Setting and Management of Fresh Water Aquarium

INTRODUCTION

The name aquarium was first used by an English naturalist Henrry Gosse in 1853 and it acquired importance very quickly all over the world, since then the aquarium has under gone a series of changes in shape and motivations. At the time of Ross Maessler 1858, fishes were assigned a secondary place but with that time, changes have taken place. Foreign small ornamental fishes and plants from the tropics and sub-tropical regions took place of our native live species.

As a result of great interest, care for the aquarium has changed from a mere hobby to scientific characteristics. Keeping a home aquarium with small tropical fishes is a growing hobby in our country. This hobby provided excellent relaxation and source of entertainment for people of all ages.

DEFINITION

Aquarium is derived from a Latin word, meaning "Water". It is an artificially prepared tank like structure in which beautiful small sized fishes live, similar to that of natural conditions.

ORIGIN

Aquarium was first originated in Japan and China some 1000 years ago. Ichthyologist collects ornamental fishes and started aquaculture. After that people of U.S.A. and European countries invented aquarium and started fish culture.

HISTORY OF AQUARIUM

First Chinese and Russian started aquariums in their gardens and houses. Then Malaysia, Singapore, U.S.A. etc. Prepared small aquariums, In India Mumbai, Mysore Bangalore, Kolkata, aquarium are made, in which ornamental fishes are cultured.

Aquarium

It is a container made up of glass or with glass walls, permits vision easily for prolonged period of time. Aquatic animals and plants which inhibit, it cared for breeding and feeding. In this, one generally strives for an aesthetic effect as well as for the most perfect possible imitation of nature.

TYPES OF AQUARIUM

1. Glass Aquaria: This type of aquaria is rounded, oval, circular, cylindrical or ball shaped.
2. Keith Glass: They are rectangular aquaria, fitted with a special type of glass (non-metal).
3. Transparent Plastic Aquaria: It is made up of plastic sheets around which metallic rod is fitted. But this type of aquaria is not durable and gets distracted soon.
4. Walled shaped Aquaria: This type of aquaria is fitted on walls. A good quality glass is used to fit around it. A particular shape of glass is fixed on metallic rod, sometimes glasses are fitted on three sides and the fourth side is covered with smooth decorative wall.

NATURE OF AQUARIA

1. *Fresh water Aquaria:* They are land aquarium and fresh water fishes are displayed in it.
2. *Marine water aquaria:* It is a complex and very tough to change the marine water to natural form. It is difficult for fishes to survive more under such climatic condition.
3. *Tropical Aquaria:* It is again of a complex type, exotic fishes are kept in it. The maintenance of these fishes is very complex. These fishes need similar conditions in which they survive under natural climate and it is tough to prepare an artificial climatic aquarium.

SPACE AND SETTING OF AQUARIUM TANK

When fish is introduced in aquarium then it cannot be shifted. After setting weed, the aquarium is placed in good climatic conditions. Thus, small room is required for glass aquarium and large room for walled aquarium.

Size of Aquaria

In making of fresh water aquarium tank, the choice of tank, depends largely on, where it is to be housed. The extent of one's enthusiasm, the money one wants to spend, there is a globe of glass aquaria of 12"-15" diameter of mouth. The glass rectangular actuaries may be obtained in a variety of sizes from 1"x5"x5" at about Rs. 25/- each to 24"x12"x12" at about Rs. 200/- each. These are usually made up of quarter inch thick clear plate glass with ground and polished edges and there is no distortion and there are no metal frames to obscure the vision, the whole aquarium is in direct vision. The great drawback of all glass tank is, that if it cracks or breaks, the whole set-up becomes useless, moreover if a larger tank is required the expenses are great.

The big aquarium is more interesting than that of "Indoor Ponds" because greater variety of plants and fishes can be kept in it, there are water lilies to suit them and generally the whole is nearer to nature such an aquarium required considerable attention. It should have plated glass sides, flat bottom and leak proof. The setting up of larger or smaller tanks is just a matter of proportion.

Location of Aquaria

After the selection of tank, a place for it is to be selected by considering its weight remembering that it will weigh heavily. When full of water a strong table will serve for its stands and this should be placed in a lighted corner with direct sunlight but not for much longer duration.

In many cases, a spare room, or shed is the best place for the preliminary operations in setting up the tank, water, aquarium should not be kept in window to avoid light penetration, algae formation due to large light penetration in it. Direct sunlight increases water temperatures which in turn is harmful for fishes, wind velocity also disturbs the movement of fishes, aquarium should not be kept in the corner of a room or far away from sun-light because it effect the rate of photosynthesis, that is why aquarium should be kept in a place where sufficient diffused light easily pours in.

Selection of Stones Gravels and Plants

The next phase of aquaria is to select small stones and gravels, which are supposed to be kept at the bottom of aquarium. Some plants are better oxygenates than others like Sagittaria can be planted 10 cm. from the back and on its either side Vallisneria and Ludwigia may be planted *Myriophyllum*, *Ceratophyllum* can be sued in clumps. Some floating plants are also used Fontinalis (Willow moss) attracts the small pieces of detritus that float about in water. The aquarium plants are of different types.

Branched Plant

- These plants have beautiful soft stem and green leaves eq. *Codia*.
- *Cabomba caroliniana* plants are branches with narrow leaves.

Rooted Plants

- Eel grass has narrow leaflets and heavy roots. Indian iron grass has greenish and brownish leaves as for example.
- *Cystoid mites* plant is a coloured plant with long stem and razor like leaves.

- Amazon sword is colorful with long leaves. In *Erracted* plant, the shape of leaves is ornamental and it is heavy rooted plant.

Floating Plants

- These plants float on the surface of water eq. Albania has long leaves and roots.
- Crystal worth has small leaves seen on the surface of water in a small rounded shape.
- Other aquatic plants are tape grass, Sagittaria, hydrophila and milfoil.

As soon as the plants are seen to be growing, a dozen snails should be in introduced, they will spend their time tidying the tank so that the fish to arrive have a nice clean home. This should be done for at least a fortnight after the tank has been planted because the plants are not well established, the fish in this exuberance will soon root them up.

Washing and Platting of Sand

Wash the aquaria to remove the impurities. Warm the sand for half an hour and a layer of warm sand is placed all-round the bottom edge of tank, this destroys the bacteria. Wash with $KMNO_4$ and dry it in sunlight. Over the top layer of sand a thin layer of stone clips or very small pebbles may be scattered. The layer of sand is about 6 cm. thick at the back of the tank and about 4.5 cm. thick in the front or the thickness of sand depends on the height of aquarium. The reason for this slope is that any debris will automatically fall to the front and thus easier to remove.

Clearing the Aquarium

Generally aquarium is washed with water or ammonia. If aquarium becomes dirty than soap water $KMNO_4$ is used, again wash it with natural water and dry it in sunlight or dry in high velocity bulb.

Oxygen and Aeration

Fishes breed in oxygenated water which is low in deep water and more on surface. Fishes come to the surface of water to inhale oxygen and release carbon dioxide. Plants and animals liberate carbon dioxide at night which is harmful for fishes. Pure oxygen is supplied by artificial aeration, which balances carbon dioxide and oxygen in aquarium, thus it forms an important relationship between plants and fishes.

Filter and Filtration

Purification of water is important to remove wastes food and decayed plant leaves. There are two method to filter water.

Mechanical Filtration

Pump is used to dispose suspended waste material. Pipe is fixed to the bottoms of aquarium and to other tank, which has a filtration chamber. The filtrated water thus collected is again transferred into the aquarium. Filtration is done by electric water pump.

Biological Filtration

Gravels are used in this process conjugated and poros, asbestos acrylic plate can be used, which is fixed on one side of the tank. Air enters through a tube, the waste product is collected on the plate and water on the other side of tank.

Lighting

Sunlight is very important for the maintenance of aquarium. Plant use sunlight for photosynthesis. In the absence of light plants will start decaying. Generally an artificial light is used in white or yellow colour which in equivalent to sunlight. High voltage light should never be used because it results in the evaporation of water.

Temperature Maintenance of Aquarium

To maintain the temperature of water in night, arrange the light which is correlated to each species of fish. Generally temperature of the aquarium should be kept at about 60°C

Selection of Fishes

Do not introduce too many fishes into the aquarium. Always introduce healthy and good quality of fishes. Selection of fish is also important. Use nocturnal species which are attractive with beautiful colours. In the aquarium six fishes of 5 cm. long not including tail may be housed comfortably. It is not wise to start with very large fish, they will knock the place about and will grow even larger in small space of the aquarium. If some are much larger than others, the small members will suffer as gold fish, mirror carp, Prussian carp, common carp, rudd, tench etc. are quite friendly with each other. The following fishes can easily live together.

Exotic Fishes

Gold fish, *Trichogaster*, angel, tartas, tacto and *Gambusia affinis*.

Mixed Fishes

Cariacus aeratus, black molly, gold fish, Trucha, *Trichogaster trichopterus*, ghost capked, black cliddo kissing gaurami, *Danio malabaricus Angel peacock* tailed, Orlando, *Betta splendens*, *Tilapia macrocephala*, *Macropodus opercularis*, *D. Symphysodon discus*,

Colias Latia and other fishes.

MANAGEMENT OF AQUARIUM

Feeding

Dried Food

Dried food spread on the surface, absorbs water and fall down to the bottom of aquarium. Fishes eat this fallen food at night, which is sufficient for the whole night.

Frozen Food

This artificially prepared food have high nutritional value.

Live Food

It includes Diafnia, small worms, insects larvae etc. Food which is not used by fishes become harmful as that food after some time becomes rotten and if such food is consumed, the fish will suffer from many diseases. Therefore, it is very important to remove all dead plants and used food from aquarium, because it affects badly on fish and plants. Instruments can be devised for removing dead vegetation or pinched off at once.

Change of Water

Water should be changed siphoning.

Removal of Algae

Algae deposits on the wall of aquarium is scrubbed by blade.

Temperature

Temperature of aquarium water should be 75°C, sometime atmospheric temperature becomes low so the aquarium temperature also decreases. To maintain constant temperature artificial heat is to be provided.

Placing of Aerator

The main function of aerator is to remove the carbon dioxide from water to outside and supply oxygen for fishes. Before placing the fish in the aquarium check the aerator twice or more if carbon dioxide increases in water than it is necessary to aerate the water within 24 hours.

Illumination

Do not use coloured bulb & and test tubes in aquarium because it does not increase the temperature of water instead increases the growth of algae.

Fish Rementation from Diseases

Remove any fungal infected or skin parasitic diseased fishes immediately because these diseases spread quickly from one another.

Changing of Plant

Dirty or rotten plants should be removed quickly from aquarium. Plants should be always shiny and healthy.

Rout in Check

1. Always check the pH of water surface.
2. Sand should be clear and away from green algae.
3. Fishes look beautiful and healthy.
4. Fishes should be active in movement.
5. The temperature should be recorded regularly.
6. Remove the waste product and algae time to time.
7. Use aerator at night.
8. Do not use light for long time.
9. Aquarium should be always covered.
10. Height of the aquarium should be proportionate.
11. Aquarium should be in the centre of room to get sufficient light and temperature.

MAINTENANCE OF AQUARIUM

Aquarium is not emptied at a time, instead its water level is maintained by pouring little amount of it as the level decreases.

The dead plants, uneaten food and other matter should be removed as soon as possible. In any case hardly ten minutes spent once in attending will solve the matter. Once in a month drop a pinch of Epsom salt into the water. The fish will gobble at the crystals as they sink. This will do them good. The crystals will dissolve and improve the water. If a plant suffers from a disease than sooner it should be removed and washed in a week solution of potassium permanganate and the tank is subjected to running water for an hour or so to wash lurking spores away.

CONCLUSION

Aquarium help in increasing the home beauty. They attracts children and provide healthy living atmosphere.

Chapter 17

Deep Sea Fishes

INTRODUCTION

The oceans covering nearly 71% of earth's surface are classified into several depth zones. There are two major zones viz, the pelagic and the benthic zones. In the pelagic zone the organisms, float or swim freely on the surface and in the benthic zone they associate to the bottom. Pelagic zone further comprises of the neritic zone, where water extends off shore to a depth of about 200 m over continental shelf and the oceanic zone that extends down vertically beyond the continental shelf to the bottom of the ocean *i.e.* water more than 200 m in depth.

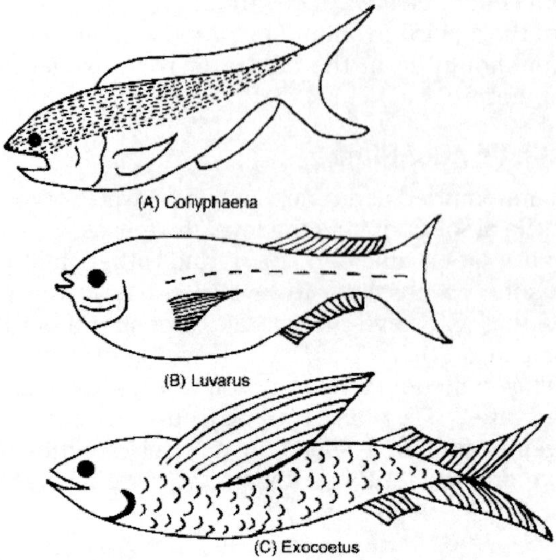

Fig. 17.1a: Epipelagic fishes

Oceanic zone is also subdivided vertically into epipelagic zone extending from surface to 200 m in sea with quite good light and oxygen in water, mesopelagic zone extending from 20 to 100 m where the light gradually fades and the temperature is in between 4-8°C. The amount of available food and dissolved oxygen is reduced, while the temperature increases; bathypelagic zones is below 100m and the abyssopelagic zone constitute the deep sea habitat which is a vast aquatic areas containing diverse and peculiar fish fauna (Fig. 17.1 a-e).

The fishes living in the deep waters of the oceans are exposed to more or less uniform conditions of life yet they are adapted differently. Various specializations are seen in their bodily form and their physiology related to the physical characteristics of deep sea zone.

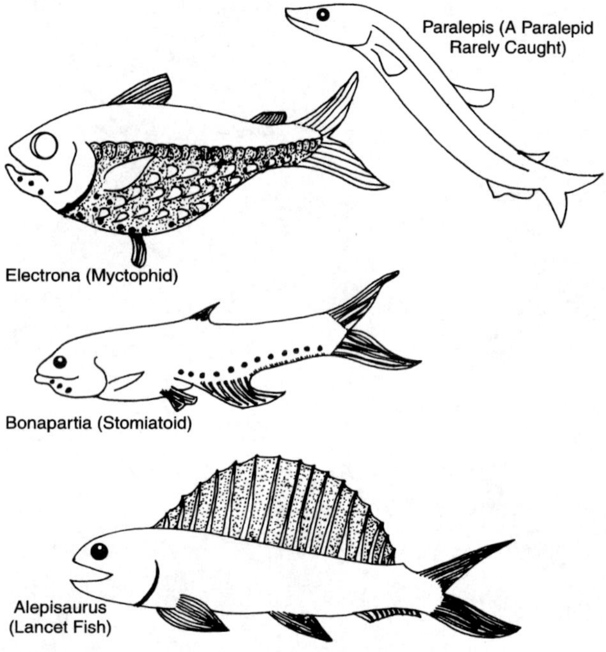

Fig. 17.1b: Mesopelagic Fishes

PHYSICAL CONDITOINS OF DEEP SEA ZONE

The deep sea life conditions are constant in time and space. These described shortly under following heads.

1. Water Waves

The waves in the ocean are restricted to the shoreward zone of the deep sea. The water remains calm and there are no water currents or wave actions.

2. Water Pressure

The pressure at sea surface measures about 15 lbs. Per square inch and the pressure increases by one ton per square inch for every 1000 fathoms and increases with the depth of the sea. The specialization of deep sea fishes are mainly in response to tremendous amount of pressure under which they live.

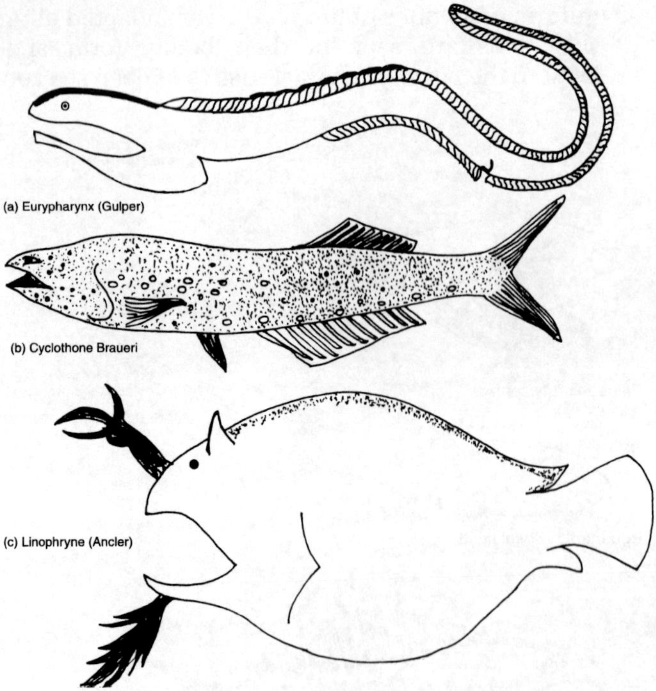

Fig. 17.1c: Bathypelagic Fishes

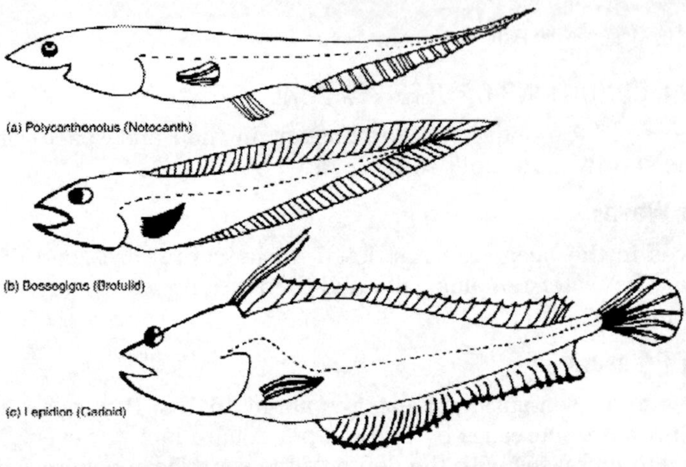

Fig. 17.1d: Benthopelagic Fishes

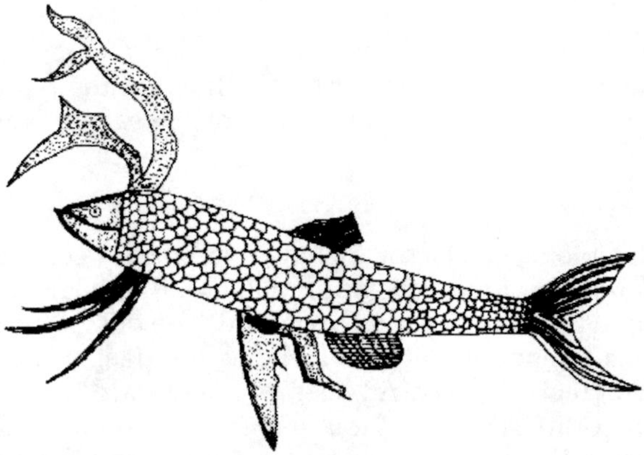

Fig.17.1e: Benthic tripod fish Bathypterois bigelowii

3. Sunlight

The sunlight penetrates the water to a depth of 200 meter and gets totally extinguished by 1000 meter in the clearest on oceans. Thus neritic zone is well lighted but the pelagic zone is lighted only at the surface. The light less zone, that plays a major role in shaping the form of the inhabiting fish.

4. Dissolved Oxygen

The oxygen content in deep sea water is quite rich. Many forms multiply oxygen in blood and others adapt to secrete it in air bladder for maintaining buoyancy. The absence of swim bladder in many deep sea fishes is related to the energy coats of filling. The bladder at great depths. Mesopelagic fishes that undertake the vertical migrations extend some energy in deflating and inflating the swim bladder, while moving up and down.

5. Temperature

The temperature of water gradually decreases with the increase in the depth of sea. Thus at sea surface the temperature is 20°C to 25°C at a depth of 200 m and at 1000 meters the temperatures is 0.5°C and it decreases from 5° to 1°C in benthic pelagic and still less in abyssopelagic zone.

6. Availability of Food

At there remains scarcity of food in the benthic zone of the ocean, there is no photosynthetic activity. Thus the forms dwelling in various zones have to depend upon the detritus falling down from the photic zone down consumed by the organisms of intermediate zone.

7. Salinity

Salinity is the main chemical factor that influences the distribution of fishes. The sea water does not show variations is salinity and does not make any differences.

8. Biological Factors

The important biological factor includes predation, competition and scarcity of food. The food consist of plants and animals falling from above and there is always a shortage of food. The competition for food is, therefore an important factor influencing the deep sea community. This results in great diversity. All deep sea fishes are carnivorous and have developed different structural adaptations to catch the prey at low temperature. Dead animals do not decompose in the deep sea.

9. Substratum

The substratum is soft and muddy. There are two factors influencing the distribution of benthic fishes, mountain ranges and ridges under the sea also form barriers to fish movement.

10. Depths

At great depths, pressure acts as a limiting factor in the distribution of fishes. The water of the ocean is relatively calm at greater depths, and there are no waves or storms, except the oceanic currents. (Fig. 17.2).

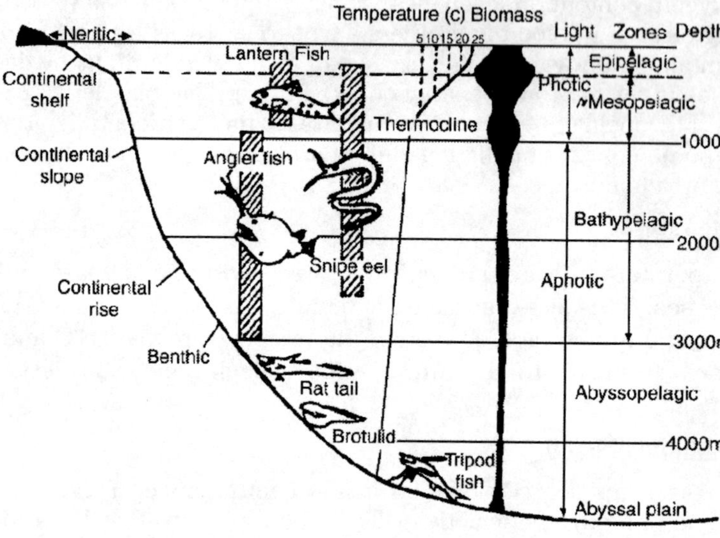

Fig. 17.2: Showing life zones and various physical condition of ocean

MODIFICATION OF DEEP SEA FISHES

General Adaptations in Deep Sea Fishes

1. Size and Weight

The size and weight of the deep fishes also decreases with the increase in depth. This relates to ambient pressure of water column availability of food materials and to the problem of sinking. By reducing the size a less or part of body is exposed to ambient pressure of working on it and by reducing their weight, the total density of the fish is reduced to counter the gravity. Weight reduction is also marked in the axial skeleton and red muscles amounting to the activity of the fish. The fishes living in deeper waters have less or amount of red muscles and they are less active.

2. Abundance

The number and diversity of deep sea fishes decrease with the increase in depth as the quantity of food gradually diminishes towards the bottom. The part of sea lying under high surface productivity zone contains larger number of big sized fishes than the parts underlying the less productive zone.

3. Buoyancy Regulation

The deep sea fishes are classified into migratory and non-migratory forms. The migratory forms migrate to the surface in search of food and experience a difference of pressure at various depths. On the other hand non-migratory forms do not migrate and confine floating into the zones to which they belong. Regulation of buoyancy and moving to different depths in water require large amount of energy which the deep sea fishes would not be able to spend. The type of buoyancy called natural buoyancy does not allow fish to sink in sea water. It is achieved by one or more of the following ways:

1. By reducing the size of the body.
2. Increasing the water content of the body brings their density nearer to that as of supporting water and reduce their rate of sinking. Example the Gempylidae have small subcutaneous water space and Tetragonus have analogous system of injecting water during its downward and upward migration.
3. Some fishes attain neutral buoyancy by storing low density lipids. The common triglycerides are stored in different parts of the body.
4. Many Mesopelagic teleosts use wax esters to attain neutral buoyancy.
5. By deflating or inflating the swim bladder.

In the primitive condition the pneumatic duct connects the bladder with the gut and the air may be allowed to get in or out of the oral cavity. The gas gland found in advanced physoclistes, do not have the duct, but secrete the gases from the bladder. The natural buoyancy is maintained by the swim bladder having a constant capacity for storing air the natural buoyancy.

4. Bioluminescence

Bioluminescence is quite widely spread in deep sea fishes which produce light by means of luminescent organs as in lantern fish and hatched fish. This is an adaptation to the absence of light. Bioluminescence organs may be present on whole of the body or may be localized in different regions as on the head or lateral sides of the belly etc.

Blind fish Ipoh's has a pair of large luminous organs beneath the transparent superficial bones on the roof of the skull. Certain sharks (Squalidae) possess numerous tiny light producing organs scattered all over the skin. Photoblepharon has light organs under the eyes. The light produced by fishes are of varied colours, used for different purposes. Malacosteus and Pachystomias produce red light.

Bioluminescence performs two functions.

1. To attract the prey as in photosynthesis Angler fish.
2. Recognition of sex and species.

5. Eyes

Eyes are very well developed in some deep fishes. They have greatly enlarged and telescopic eyes. The eyes of fishes enable to collect as many rays of light as possible. Telescopic eyes may be directed forwards as in *Gigantara* or upwards as in *Argyropelecus*.

In Mesopelagic fishes, the eyes are usually larger and adapted to see in dim light. The eyes have large pupils to absorb maximum amount of light, Retina has a high concentration of visual pigments and an increased number of rods to accommodate them. It is capable of absorbing 95% of light. Many other mesopelagic fishes having tube like eyes to enlarge the field of vision.

In *Gigantara* the tube shaped eyes point forward to look in front upwards, but the eyes of Argyropetecus can look upwards, when the fish rest obliquely at the bottom.

Certain fishes having eyes which are more like the concave mirrors to maximize the amount of light. Bathypelagic fishes living in the complete darkness of the sea having small eyes which vary in size and helps in detection of bioluminescence and recognition of sex and species. Their reduced size is related to the limited energy, which the fish can manage to spend.

Many deep sea fishes are completely blind but they use a well-developed system of lateral line canals, the tactile organs and other sensory structure to compensate the loss of vision, while in some fishes olfactory organs are very well developed.

6. Food and Feeding Habits

In deep sea, the food material consists of dead animals and plants falling like a gentle rain. But the quantity of food diminishes gradually with the increase in depth because of column feeders. The deep sea fishes are predatory in habit. The scarcity of food is a tendency towards the strict carnivorous habit and conservation of energy increase with oceanic depth. Deep sea fishes evolve provisions to spend less energy while sharing their prey.

In Mesopelagic zone, the fishes, (*e.g.* myctophids and stomiatoid) have vertical migration to the surface to feed upon the zooplanktons. According to Marshall (1979) feeding migration is common in fishes of this zone. Some fishes are predatory having large powerful jaws, with strong teeth and long alimentary canal for consuming the prey larger than their size *e.g.* Alepisauroids and chiasmodontoids. Bathypelagic fishes live in scarcity of food and have a limited budget of energy, so reduced organ system.

Deep sea fishes have well developed jaws, wide mouth, strong dentitions, well developed gill rackers and long stomach *e.g. saccopharynx* (Gulpered). *Chiasmus melanocytes* can consume prey two or three times larger than their size. Deglutition is effected in these fishes have well developed jaws, wide mouth, strong dentition, well developed gill rackers and long stomach *e.g.* saccopharynx (Gulpered). *Chiasmus melanocytes* can consume prey two or three times larger than their size. Deglutition is effected in these fishes by the action of jaws as in snakes and not by means of the muscles of the pharynx. Benthopelagic fishes depend mostly upon the Benthopelagic zooplanktons. Many of them feed upon the decay in goose and have lost the power of mastication. The Bathypterois possess stiff fin rays for balancing while if feeds upon the copepods and other crustaceans.

7. Other Sense Organs in Deep Sea Fishes

Some deep sea fishes have no sight but have well developed long feelers that keep exploring the surroundings continuously. In Bathypterois a single ray of various fins are longer than the body, acting as feelers. *Stylophorus paradoxus* has a long filament, arising from the caudal fin. The long feelers of deep sea fishes enable to perceive the slightest movement of water in their neighbourhood. They are always afraid of attacked by the enemy. As their lateral fine system is very well developed they remain aware of even slight disturbance in the surrounding water. (Fig. 17.3).

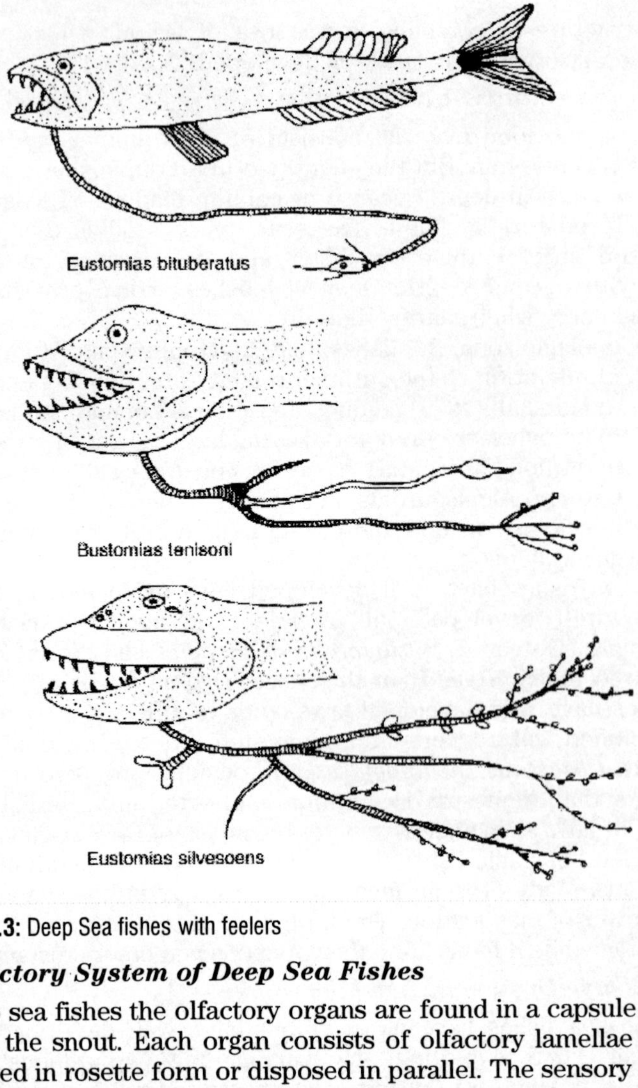

Fig. 17.3: Deep Sea fishes with feelers

8. Olfactory System of Deep Sea Fishes

In deep sea fishes the olfactory organs are found in a capsule on each side of the snout. Each organ consists of olfactory lamellae that are fashioned in rosette form or disposed in parallel. The sensory cells are innervated through axon that run back to glomerular synapses in olfactory bulb, which is connected to the fore brain. In deep sea fishes as in other fishes, the olfactory organs are diversely developed. At one extreme the organs are large (Macromatic) as in the sharks and bottom-dwelling eels and on the other they are small (Microsomatic), as in female angler-fishes (Ceratoids). Fishes with the best developed sense of small have the most numerous olfactory lamellae. The fresh water eel (Anguilla) which spends part of life as a deep sea fish, is a good example.

9. The Chemical Sense, Smell and Taste of Deep Sea Fishes

Fishes may have three other chemosensory means single taste cells over the bodies and perhaps present in the mouth and free nerve endings in the skin and receptors in the lateral line system.

In deep sea fishes taste is restricted primarily to aspects of food finding and feeding, but smell while they search for food detection and location, in reproductive and social behaviour (Bardach and Villars, 1974).

10. Colours of Deep Sea Fishes

The deep sea fishes are dark coloured as cyclothene whale fishes, gulper eels, the Angler fishes and frog fishes from most of the fish fauna. The most fishes are black or silvery with counter shading as a means of concealment and protection from the predators.

11. Reproductive Strategy of Deep Sea Fishes

Deep sea fishes show remarkable adaptive features for reproduction. These fishes face the problem of finding a mate due to the absence of light. The remarkable specific bioluminescence and the sex pheromones produced by the females check are sensed by the olfactory organs of males. The large organs are capable to find the odour of their mate. Some deep sea benthic fishes have solved the problem of finding a mate by producing sounds with their swim bladder as in Macrouridae and ophidiidae. Photophores also serve to attract the mate. In many species of bathdy pelagic fishes, the male is smaller to conserve food and energy but in some species, the males do not feed although they are fee living. Another adaptive development to find a mate is to become hermaphrodite so that any individual can mate with any other individual, as in deep sea predatory lancet fishes and barracuda as. In some species the male after finding the female becomes attached to her by means of jaws and assumes a parasitic life. In such species the gonads of both the sexes do not mature until the female has been parasitized by the male.

Some species have developed parental care and a few are viviparous. Thus various strategies have been evolved by the deep sea fishes to ensure reproduction to maintain the population.

12. Life Cycle in Deep Sea Fishes

a. Eggs

Deep sea fishes spawn at a depth in which they live, but their eggs always rise to the surface as they are neutrally buoyant. The eggs are produced in large number which increases the chances of their survival. The eggs of *Viviparous squaloids* contain squalene that makes them neutrally buoyant. The large number of epipelagic eggs are produced *e.g.* female angler fish may produce upto 5 million eggs.

b. Larvae

The larvae hatch in surface water where enough food is available for the larvae and the conditions are favourable for the development of the eggs of deep sea anglers that hatch into the common kind of fish larvae. The newly hatched larvae are globular and short tailed sexes are distinguable at this stage as they feed and grow in the upper waters.

The larvae are generally transparent to escape from their enemies. After metamorphosis the young ones move down to the habitat of their parents. The skin swells like balloons to reduce density and the rate of sinking.

Chapter 18

Fish Diseasse and their Control

INTRODUCTION

The fishes are obtained through capture and culture methods, but finding diseases free fish of good quality may be a problem at times.

When certain abnormal changes in the body like structural functional or behavioral and produced, that characterizes of certain disease and nutritional deficiency or injury. Large scale mortality of fish often occurs in ponds and tanks due to parasitic infections like fungal, bacterial and protozoan. Water is mainly responsible for various diseases and the vital functions of fish movement, feeding, digestion, assimilation, growth, responses to stimuli and reproduction are dependent on water.

In tropical countries like India, the volume of water in a pond fluctuates very much in relation to seasons. During summer a reduction in the living space in the ponds due to evaporation may make fish crowded. The perennial ponds are never dried. However, year after year, the organic matter is accumulated along with the spores and eggs of the parasites, which in the water causing wider chances for quick infections. Then food originated diseases and other diseases like Asphyxiation, physical and chemical diseases were observed in fishes.

These pathological changes are clear deviations from the normal state, anatomically and physiologically and are known as pathological changes. Many of these are basically protective. Mechanism arising in response to the hostile stresses. The study of the distribution and determinants of a disease or an infection in the host or in the vector species is termed as Epizootiology (Roberts 1978) Aetiology on the other hand deals with assignment of cause for the disease.

Principal Changes with Cell Death

There are three principal types of necrotic changes with cell death

a) The enzyme are released either from the leucocytes (neutrophils) or from the pathogens (bacterial toxin) due to enzymatic lysis.
b) Coagulative in nature, resulting from loss of blood supply to the tissue, (ischaemia) the nuclei are destroyed completely.
c) Affecting nucleus which may become shrunken and dark (pyknosis) or show break down of nuclear membrane coupled with chromatin fragmentation (Karyorrhexis) or loss of basophilia

following hydrolysis of the nucleic acid followed by irregular nuclear profiles *i.e.* sickle-shaped or tadpole-like, with chromatin strands persisting (Karyolysis).

PATHOLOGICAL CHANGES IN FISH ORGANS

Some pathological changes in fish are enumerated below:

Spongiosis of Epidermis

This is an inflammatory response and individual cells in the malpighian layer becomes separated by tissue fluid. That is known as intercellular oedema. This conditions reflects a pathological lesion in the dermis. The epidermal cells die following necrosis of cytoplasm and nucleus.

Hyperplasia of Epidermis

The epidermal cells proliferate at all levels.

Ulceration of Epidermis

The epidermis tends to get separated from the basement membrane. This is caused by necrosis of adjacent dermis.

Dermatitis

The upper layer of skin (*Stratum spongiosum*) and lower layer of skin (hypodermis) are commonly affected. The middle layer stratum compactum is not affected. The inflammatory response is due to bacterial infection. Vascular infiltration is conspicuous. Exudates contain necrotic debris, fibrin and pathogens.

Anaemia

Anaemia is due to the reduction in the total amount of haemoglobin carrying red blood corpuscles, leading consequently to reduction in oxygen transport capacity of blood and is caused by pathogens attacking haemopoietic tissue.

Leukaemia

The number of leucocytes increases in the blood.

Neoplasia

The formation of tumours in the affected tissue caused by a number of factors, such as virus, toxin, hormones etc. The common tumours in fish include papilloma of skin, carcinoma of oral mucosa, including lips, adenoma of thyroid and liver fibroma of connective tissue, lipoma of fibrous tissue, leiomyoma of smooth muscles, rhabdomyoma of striated muscles chondroma of cartilage and osteoma of bone.

Oedema of Gills

The base of secondary lamellae shows the increased capillary permeability. There is an extensive oedematous separation of the gill epithelial lining which is followed by cellular necrosis in the epithelium.

Granuloma of Gut

Small white nodules appear on the gut wall in chronic infection.

Cirrhosis of Liver

Toxic agents cause extensive cell necrosis. The nuclei pyknotic and the cytoplasm vacuolated. Cirrhosis which represents an advanced condition as hepatocellular damage. Fibrosis occurs specially in the peribiliary connective tissue, virus diseases produce focal necrosis.

Glomerulonephritis of Kidney

Thickening appears on the Bowman's capsule's glomerular tuft and also on the glomerular basement membrane. Fibrosis may follow, because of the diabetic conditions and parasitic or bacterial infection.

Necrosis in the Whirling Disease of the Ear

Pollution and myxosoma cerebralis produce necrosis of bone and cartilage resulting in deformation of the canals. This causes loss of balance control in fishes.

Pathological Changes in Lateral Line Organs

Pollutants like hydrocarbons, pesticides, detergents, heavy metals etc. produce a wide variety of pathological changes including oedema, spongiosis, necrosis sloughing and hyper and metaplasia of sensory epithelium.

Cataract of the Eye

This is common among farm fishes caused by the parasite *Diplostomum spathecum*) However, cataract may be caused by a wide variety of other factors including diabetes.

Lesions of Muscle

Muscle is favourable for different parasites, which produces different types of lesions, example lernean (Crustacean) causes liquefactive necrosis. Cryptocotyle causes fibrosis and nematodes causes granulomata.

Haemorrhage in Testis

This is commonly excited in tuberculosis and furunculosis.

Sterility of Ovary

Dietary deficiency and diseases are causes of pathological changes. Artesia (failure to produce primary oocytes) is one. Reabsorption of mature ova is brought about by phagocytic activity of macrophages in the event of failure to mating. Pleistophora ovariae excites necrosis, cyst formation and fibrosis in the ovary of Notemigonus a cultivated bait fish ligula and schistocephalus infestation went heavy exerts pressure from abdomen to prevent ovarian growth.

In many parasites and pathogens, the infection is direct. The agents freely swim on the host by contiguity such as fish louse (Argulus), leach (Piscicola) and aquatic fungi. In other cases it may be through a vector *e.g.* the fish leech (*piscicola geometra*), transmitting trypanosoma in cyprinids, causing the sleeping sickness (Trypanoplasmosis) in the host fish. In other cases, the infection may be caused through the medium of food which is already infected. For instance, the transmission of acanthocephalan (*Echinorhynchus trutta*) through the amphipod food in trout and salmons feeding upon them or transmission of cestode (Ligula) through copepods harboring the larval stages of the parasite when a fish eats these copepods.

PROCEDURES FOR PATHOLOGICAL CHANGES OF FISH DISEASE AND DIAGONOSIS

Following procedure are found to be useful.

Assessment of Fish Environment

It is easier to examine health condition of fish in water with greater transparency and lesser depth.

The fish pond of cold water culture is beneficial for direct examination but the fish pond of warm water culture do not permit direct examination because of lowered transparency and greater depth. It is necessary to examine a sample fish farm, time to time.

Assessment of Health of Fish

A sick fish differs from healthy one in many ways, particularly in behaviour. Common test include escape reflex defensive reflex, ocular, reflex, tail reflex etc which are much changed in sick fish.

i) Sick fish is less timid to external disturbance or not at all.
ii) The sick fish may be calm and sluggish and not violent when taken out of water.
iii) The sick fish when held in hand by head, its trunk and tail hang down vertically with fins folded and don't remain horizontal with fins spread out.
iv) When held with back facing upwards and rotated, sick fish makes

no effort to keep its eye balls remain in original position, while a healthy fish in variably does when it is rotated.

Assessment of Possible Causes of Sickness

1. By evaluation of quality of water in which the fish is present or an epidemic breakout. Aliments due to poisoning or suffocation or mortality, may develop with changes in the water quality which causes, water to smell, transparency, colour, oxygen content, temperature, pollutant etc.
2. Closely examine the sick fish for diagnosis of the disease and its causative factor. Lucky (1977) on carps gave following account:
   a) The fishes come to surface and show little escape, reflex then sink to the bottom and lie there waiting for death a kind of chemical poisoning.
   b) The fishes come to surface and gulp air bubbles frequently. After some time the fish starts dying with mouth wide open and opercula spread apart a kind of asphyxiation due to low oxygen concentration or damage of gill due to parasitic disease.
   c) Carps appear sick especially of one species and of one age group come to float on the surface. Examination of sample fish reveals some organ or organs of body that are affected and Aliment is generally prolonged.

Greyish blue marks on skin	-	Gyrodactylosis.
Small boils on skin, gills	-	Myxobolosis.
Cataract of eye	-	Diplopstomiasis.
The white spots on skin and gills	-	Itchy phthiriasis.
Cotton tufts on skin & gills	-	Saprolegniasis.
Grey regions on gills	-	Branchiomycosis.
Ulceration of skin, skin lost at places.	-	Dropsy.
Presence of lice on skin	-	Argulosis.
Presence of fish leeches on skin	-	Piscicolosis.
Internal organs affected	-	
Wounds with sharp margins	-	Injury from attacks of birds.

Major Symptoms and Diagnosis of Parasite Infections and other Diseases

Fish Scraps Sides

a) Scrapings show trematodes.
b) White spots on skin which show ciliated protozoans.
c) Grayish film on skin and fins with loss of appetite, on set of

debility, skin mucous film shows active protozoa.

Fungus

The body surface shows tufts of gray filaments.

Fin Rot or Tail Rot

The progressive disintegration of the tissue of the fin first starting as a white line at the outer margin.

i) Rod shaped bacteria seen under high power microscope.
ii) Blisters and ulcers appearing on the body and causing sudden mortality. Haemorrhage occurs on swim bladder, large intestine etc. Autopsy of kidney under microscope reveals rod shaped bacteria which are the cause of furunculosis.

Lice

Tiny copepods clinging on gills and body.

Grubs

Cysts appear generally of a size of a pin head, occurs on various body parts of fish. Viz scales, flesh, liver, kidney, fin, spleen integument gonads, lens other parts of eye, gills, etc. The parasite may be tapeworm or flatworm or roundworm.

The examination of affected organs requires experiences and skill. Some laboratory procedures are

a) Removal of fluid contents from inside the boils, knots, vesicles, abscesses etc.
b) Collection of blood sample of diseased fish for haematological examination.
c) Culture of exudates from affected internal organs of the fish body.
d) Preparation of slide-smears from diseased organs or from cultures grown from their exudates.

TYPES OF DISEASES

The disease related to sanitary conditions prevalent in water are as follows

Parasitic Infection

Fungal, Bacteria, viral, protozoan, worms, Arthropods, Annelids, and mollusc.

Non Parasitic

Asphyxiation, Physico-Chemical condition of water.

Miscellaneous Diseases

- Pox diseases, gas bubble disease algal blooms.

- Nutritional diseases
- Disorders by Abiotic factors
- Disorders by biotic factors
- Acidosis and Alkalosis
- Deficiency of oxygen

Fungal Disease

These are among the commonest disease of fresh water fishes, fish fry and fingerlings. The adults (major carps) get mechanically injured during transport. The fungal diseases are characterized by the appearance of tufts **"Cotton wool"** on the skin (dermatomycosis). This mould spreads very rapidly. Two types of moulds are common. The mould (Saprolegnia) causes gill not (branchiomyces) of carps. The infected gill begins to rot. The colour of gill first changes to pale and then shows red patches. In advanced stages the gills change yellow brown and are partially destroyed. The carps of age group of 1-2 year are easily infected. The other fungal disease is the general saprolegnia infection in fresh water fishes, living in organic rich water. The fish contract infection when it bruises itself. Symptoms include small white tufts on fin and skin, later on gills, become affected. The infection spreads quickly in an injured, weakened or diseased fish.

Bacteria Diseases

The common bacterial diseases are ulcer, dropsy, eye disease and fin or tail rot. But these are most feared diseases causing heavy loss to fish farms. Two are very important.

i) **Dropsy:** It is serious infection in major carps. The bacterium *Aeromonas punctata* is responsible for the disease. It is very common during spring in temperate climate. It includes general oedema, accumulation of watery liquid, yellow or pink in colour, in the body cavity and withering away of fins. Intestine are inflamed and liver and kidney are affected. There is haemorrhage in most tissues. Normally the bacteria remain dormant if the infected fish is in good health. Weak fishes will, however be harmed and may even die. European carps are easy victims suffering heavy losses.

ii) **Furunculosis:** It is caused by *Aeromonas salmonicida* in trout's and salmons. The fish may get infected on gills, gut and wound on surface. The bacteria remains dormant in healthy fish, weakened fish in bad conditions are harmed. Death may occur within few days following the infection of disease. The characteristic symptoms is the appearance of furuncles (boils) these are round swelling appearing on the sides on the body. Each swelling contains a pus like substances which is a mix of destroyed body tissue (muscles and leucocytes etc) and bacteria. Internally intestine shows heavy inflammatory conditions which is seen on the peritoneum and muscles.

iii) **Eye disease:** It is caused by bacterium *Aeromonas liquefaciens*. It affects eyes, optic nerve and brain. Cornea becomes red and later on becomes opaque and eye ball degenerates.

Viral Diseases

The most important is the viral hemorrhagic septicaemia (VHS). The virus contains RNA genome symptoms include general anemia, pale discolouration of gills, swelling of eyes and belly, oedema of muscles, inflammation of intestine etc. Sores may appear on skin, anus and often kidneys and liver are also affected. There is haemorrhage in most tissues. The next important viral disease is the infectious pancreatic necrosis (IPN) in salmonids. The virus contains RNA genome. The young fish is easily affected. The diseased fry looks darker and swims in a spiral fashion. Mortality is considerable. Acinar pancreatic cells show extensive necrosis followed by rupture and release of zymogen granules. Another is infections dropsy of carps called spring Viraemia of carps (SVC). The affected fish shows greatly distended abdomen, loss of balance, haemorrhage on skin and gills and darker colour. Rainbow trout suffers from an important viral disease known as viral hemorrhagic septicemias (VHS). It causes anemia. The gills becomes pale, eyes and belly swell, oedema of muscles, inflammation of intestine also occurs. Liver and kidneys are also affected. There is a hemorrhage in most tissue. The channel catfish virus (CCV) disease is caused by DNA genome. Affected fish hangs vertically in the water and shows loss of balance. There is hemorrhage of visceral organs.

Protozoan Diseases

The number of protozoans cause numerous external and internal diseases particularly in the young fish. Fry and fingerlings are their large-scale victims. Ichthyophthirius multifiliis infects different regions of the body externally causes white spot itch disease (Icthyophthriasis). The *Trichodina indica* (ciliata) infects mostly major carp. Costiasis is caused by *Costia necatrix*, and *Bodomonas rebae* (Mastigophora) produce complains in Rohu, Catla and Mrigal. Then others like leptothecae, lenthospera, myocardium are common myxosporidians fish parasites. Myxobolus (a sporozoan) causes ulceration or boil disease producing tiny knots in the skin of *Cyprinus, carpio*, Trypanosoma causes sleeping sickness by infecting blood of skates, sharks, pike, eel, rudd, common carp and perch etc.

Worm Diseases

Worms of four groups are parasitic on fish, they are flat worms (trematodes), tapeworms (cestodes) round worms (nematodes) and thorny-headed worms (acanthocephalans). The worms generally affect the internal organs, especially the intestine.

Flatworm infection is mostly caused by Distomum and Dactylogyrus. It infect gills only. The gills show swelling and marginal discolouration. The diplostomum causes Diplostomiasis or black spot diseases spot present on all over the body of fish. The tapeworm (cestodes) infection is more serious than the flatworm infection. It affects the alimentary canal and gall bladder. The ligula intestinalis infects fish, shows swelling of the belly and the gonads are adversely affected with loss of fertility. In final stage death occurs. The Nematodes (Heliconema, Proleptus, Camallanus etc) are affecting the fish. They affects alimentary canal. The Acanthocephalans *i.e. Acantho gyrus*, Pallisentis etc are endoparasite, infects fishes.

Annelidian Disease

The leeches belong to families Rhyncohdellidae and Gnathobdellidae, infect major carps and feed on the blood of the host fish. They cause irritation and abnormal movements of the host. They are ectoparasites, thus infective fishes become anaemic and weak.

Arthropodan Disease

The crustacean diseases are caused by Argulus, (Carp Lice) Lernaea, Ergasilus. These are ecto-parasites on fishes. The Argulus infection is one of the commonest infection. The affected part shows red spot on the body stunted growth, loss of scales and sucking blood. Lernaea also known as anchor worm are buried deep into the skin of the fish and only the posterior region remains visible. The Ergasilus is generally found attached to the gills of fishes and gills lamella became weak and degenerate.

Mollusc Larvae

Glochidium (Mollusc larvae) larvae cause heavy infection in fishes and result in the death of a fish. The main symptoms include the damage of gill arches, filaments and appearance of tumours.

Non Parasitic Disease

Asphyxiation, Physico-chemical condition of water.

Asphyxiation

1. Deficiency of Oxygen

Oxygen is essential for life. It is dissolved in water. Aquatic animals consume oxygen during respiration and is also utilized by the putrefaction of organic matter. Thus an imbalance is caused.

2. Physical Conditions of Water

Depth, temperature, turbidly, light etc are responsible for proper growth. The change in these causes mortality.

3. Chemical Agencies including Pollution

Different types of pollutions are responsible for the mortality of fish symptoms and treatment differ on the basis of quality and quantity of the pollutions.

4. Physico-Chemical Properties of Water

The physico-chemical properties of water are of great importance in fish culture.

5. Too much Acidic Water

When the pH of water goes below 5.5, the water becomes gradually toxic to fishes. A whitish film covers the body of the fish and mucous is secreted. Natural mortality starts.

6. Too much Alkaline Water

The pH of water should always be kept below 9.0 otherwise, it will be again toxic for the fishes. This also affects the gills and fins of the fishes.

7. Temperature Variation

Higher temperature variation is also harmful for the fishes. A difference in temperature between 10 to 12°C is well tolerated by the fish. A suitable water temperature for salmon is 10°C, for perch it is 12°C, for tilapia the range is 10-23°C, and for major carps it is between 20-37°C.

8. Lack of Oxygen

Oxygen deficiency causes suffocation in water for the fishes and mortality can start. Presence of aquatic plants should be maintained. The oxygen requirement varies in different fishes, at 20°C, salmonids require 9mg.

Miscellaneous Diseases

Pox, diseases, Gas bubble disease, Algal blooms.

1. Pox Disease

The small white milky spot present on the body of fish which gradually increases in size, later the spot becomes solid and the fish becomes weak, slow and then dies.

2. Gas Bubbles Diseases

When nitrogen of the water is higher than 125%, saturation is due to rapid temperature changes, gas bubble disease may result and fish fry die in large numbers. Fishes affected by this disease swim at an angle of 45°. The symptoms include the presence of bubbles beneath the skin, on fins, around the eyes, in stomach, intestine and blood capillaries. The nitrogen and super saturated level of oxygen have also been reported to cause gas bubbles disease in fishes.

3. Algal Blooms

In summer the Algal blooms spread all over the surface of water. It causes obstruction to the sunlight. So the percentage of dissolved oxygen is lowered. Thus, the gills are chocked causing the mortality of the fish.

4. Dysentery

In some fishes dysentery have been reported.

Nutritional Diseases

The nutritional diseases fall under three groups

a) Arising from under nutrition.
b) Arising from dietary deficiency in the major components of food.
c) Arising from toxic effect of the diet.

Valued from fishes; grown in fish farms need high quality food and thus, these develop disease from under nutrition. Standard artificial diets tend to eliminate these diseases whereas sub-standard diets in supplementary feeding increase the possibility of this incidence.

Lipoid hepatic degeneration disease is common in farm fishes that the liver takes a yellow brown colouration. Anaemia follows with gills looking paler. The liver is swollen and the hepatic cells are infiltrated with lipid. The disease is significant in salmonid caused by over feeding of pelleted diet. Vitamin 'A' imbalance (Vitaminosis A) produces varying degree of syndromes. Its deficiency causes, under growth blindness, hemorrhage at the fin base and keratomalacia. Its excess in the diet causes, squamous metaplasia, splenomegaly and hepatomegaly. Vitamin B6 (Pyriodoxine) deficiency affects early larval growth where it causes lesions in the central and peripheral nervous system. The affected fish shows such nervous signs as gasping of the opercular folds etc.

Nutritional gill disease results from pantothenic acid deficiency in the diet. Secondary gill lamellae are clubbed and primary lamellae shows hyperplasia. Inappetence, respiratory troubles and anemia, are shown. Ascorbic acid (Vitamin C) deficiency results in spinal deformities due to abnormal skeletal development and poor wound healing. Toxic components of food specially of the artificial food pose great problems in fish culture. A flat toxin develops in mould infected oil seed meals from the metabolic activity of the mould, Aspergillus flavus, Such diets cause neoplasia (hepatoma) in which hard tumours appear on the body behind the pectoral fins specially in trouts. The necrosis of renal tubules and depression of hemopoiesis occurs. This toxin damages the kidney and liver where it causes glomerulonephritis, necrosis of renal tubules and hepatic cells.

Disorders by Biotic Factors

Certain algae phytoflagellates, dinoflagellates and blue green algae due to their excessive proliferation and toxins cause various disorders in fish *Prymnesium parvum*, A gill damaging species of phytoflagellates causes mortality in tilapia, gambusia and grey mullets. Blooms of dinoflagellates, secrete neurotoxin cause large scale fish mortalities and other algae effects on skin, fins, of herbivorous fishes. They effect on internal organs where generally hemorrhage occurs.

Disorders by Abiotic Factors

The domestic and industrial effluents have harmful effects on fishes. The pollutants include ammonia nitrogenous compounds, detergents, chlorine, phenol, various salts, cyanides, insecticides, molluscicides vermicides, rodent poisons, disinfectant, herbicides etc. The impact of waste water poison in culturable fish is given below

	Substances	Disorders
1	Phenols, synthetic detergents and alkalies	Haemorrhage on skin, head, base of fin, discolouration of gills, change in blood, decrease in erythrocytes.
2.	Chlorine (<4mg/lit)	Damage of gills, anaemia and discoloured, death after 8 hrs. in carps.
3.	Iron and nitrate	Increase in erythrocytes.
4.	Arsenic salts, manganese chloride and acids.	Blindness.
5.	Ammonia (>1mg/ltr)	External bleeding and haemorrhage in internal organs.
6.	Carbon dioxide	Elevated levels causes nephrocalcinosis, calcium deposits in ureters and swelling of kidney.
7.	Hydrogen sulphide (>6mg/litre)	Gills become reddish violet.
8.	Insecticides	Liver cells and blood vessels are damaged.

Acidosis and Alkalosis

A great majority of fishes live in water which have pH-7.8. If the pH of water goes down drastically, there in a reduction of calcium salts or release of humic acids from the soil, this phenomenon is known as acidosis. The fish may show very rapid swimming movements and jump out of the water. In the gills of carp, acidosis causes dark grayish deposits, darkening the edges and mucous secretions.

Aquatic plants present in high densities liberate enormous quantities of oxygen, during photosynthesis which is responsible for the formation of insoluble calcium carbonate from calcium bicarbonate followed by the formation of calcium oxide with the elimination of carbon dioxide. This phenomenon is known as alkalosis. The alkaline condition leads to the corrosion of branchial epithelium of fins. Alkalosis can be prevented by buffering the medium by means of suitable calcification.

Deficiency of Oxygen

All gases are dissolved in water. Oxygen is most essential constituent for the life of the fish. The oxygen consumption in water takes place during respiration by the aquatic animals and also by the purification of organic matter, thus an imbalance is caused. This effects the fishes causing anaemia and fat contents are also increased.

TREATMENT AND CONTROL MEASURE

Treatment of Fungal Infections

Dip the affected fishes in 3% sodium chloride solution or in 1:2000 solution of copper sulphate or in 1:1000 potassium Permanganate solution for about 5 to 10 minutes or until fish shows distress. For gill rot, the affected fishes are kept in 3 to 5% solution of sodium chloride or in 5 ppm. Potassium permanganate solution for 5 minutes, 1 to 2 ml. of malachite green solution may be used in one litre of water for a bath lasting for about one hour.

Treatment of Bacterial Infections

The antibiotics (Chloramphenicol) may be used by infecting (1 to 1.5 mg for every 100 gm. body wt.) or by mixing with food.

1. Ulcer Disease

The infected fishes are kept in 1:2000 copper sulphate solution for about 3 days.

2. Dropsy

Dip the fishes in 5 ppm. solution of Potassium permanganate for about 2 minutes. No food should be given during treatment.

3. Eye Disease

Chloromycetin (8-10 mg/lit) bath for one hour daily for 2 or 3 days are given and the entire pond should be disinfected by 1 ppm. solution of Potassium Permanganate.

Treatment of Viral Infections

Currently there is no effective treatment, medicines are of little use. Remove the diseased or dead fish from the pond. Avoid fatty food but it should be rich in vitamins (A, B Complex, E, T) containing appropriate proteins.

Treatment of Protozoan Infections

The infected fishes are treated with 1.5000 formalin solution for an hour each day for a week. Trichodina indica infects major carps. Another disease costiasis is caused by *Costia necatrix* and *Bodomonas rebae* in catla, rohu and mrigal. The pond may be treated with 1 gm. of malachite green per 12 metre square area thrice at intervals of 2 days for Icthyophthriasis.

Treatment of Worm Annelid Infections

The infected fishes may be treated, with 1:2000 solution of formalin or salt bath for 5 minutes. The affected fishes (Small black spot) are dipped in 3:10000 picric acid solution for one hour. Leech infected fishes may be treated with 1:1000 solution of acetic acid with 1:10000 Potassium permanganate solution 5 ppm of Gammexane, cure the infected carps. Lysol bath (1ml. In 5 litres) for 15 seconds and then rinsed with water also cure fishes.

Treatment of Arthropods Infections

The fish lice Argulus, Lernea and Ergasilus are ectoparasites on fishes. The infected fishes are treated with 1:1000 solution of acetic acid for 5 minutes followed by 1% sodium chloride for one hour (Argulus). Lernaea and Ergasilus are removed mechanically. After that the infected fishes are dipped in a weak solution of Potassium Permanganate for 2-3 minutes.

Treatment of Mollusc Infections

Mechanical removal and 1:1000 solution of acetic acid for 5 minutes, followed by 1% solution of sodium chloride for one hour.

Treatment of Non-Parasitic Infection

1. Too much Acidic Water

500 kg. of calcium carbonate should be spread in one hectare of the pond so that the desired pH of water can be obtained.

2. Too much Alkaline Water

The distribution of quicklime and biogenic decalcification by controlling and reducing submerged vegetation can help in reducing the pH of water.

3. Temperature Variations

Maintenance of temperature in pond depending upon different species of fishes.

4. Lack of Oxygen

The oxygen requirement varies in different fishes to maintain the oxygen concentration in pond.

Miscellaneous Disease

1. Dysentery

Animals should be segregated and hard food should be stopped.

2. Pox Disease

Infected fishes are treated with 5% Potassium Permanganate solution every day for 5 minutes.

3. Gas Bubble Disease

In such conditions, water should be well agitated to bring down the nitrogen saturation below 110% or affected fish should be transferred to other ponds. The oxygen content should be balance in ponds.

Nutritional Disease

Thyroid tumours may develop due to the deficiency of potassium iodide. A mixture containing one part of iodine and 100 part of potassium iodide, can be added in the food in the ratio of 1:2,500 to avoid the development of tumours.

Treatment of Disorders by Biotic Factors

The affected fishes are treated with 10ppm ammonium sulphate or 2-3 ppm copper sulphate solution.

Treatment of Acidosis and Alkalosis

In the event of mortalities in ponds due to the acidosis the pH must be normalized with powered calcium carbonate and not with quicklime. Alkalosis can be prevented by buffering the medium by means of suitable calcification.

OTHER TREATMENT

1. Treatment by immersion for external chemical therapy.
2. Formalin is a traditionally tried chemical. It is very effective at

conc. of 167 to 250 mg/litre for one hour (Roberts) 1978.

Organophosphates

Organophosphates are used for ectoparasites like Lernaea and Argulus and some treatments as Gyrodactylus.

Malachite Green

Malachite green is effective against fungal infections including saprolegnia. It is used at a concentration of 1-2mg/litre for 1 hour.

Copper Sulphate

Copper Sulphate is effective against bacterial infections.

Ammonium Compound

Quaternary ammonium compound is very effective against bacterial infections. It is generally used at a concentration of 1-4 mg/litre.

Roccal

The use of Roccal (Alkyl dimethyl benzyl ammonium chloride) has been recommended. This chemical is used at a concentration of 1:50,000 and the fish may be dipped in the solution for one hour. This chemical is highly toxic to the fish and has to be used carefully.

PMA Compound

The new compound PMA (Pyridyl mercuric Acid) has been widely used. The advantage is that it is much less toxic to the fish. It has been found that a solution of 1:5000 is effective. This chemical is used for the control of external parasite.

Treatment by Drugs via Food for Internal Administration and Therapy

a) Furazolidone: A nitrogen popular used in veterinary therapy for incorporation in fish food in the case of bacterial infection. The dosage required is 11gm/100kg fish/day for about a week.

b) Sulfadiazine: A sulfonamide, mixed with trimethoprim, a potentiated sulfonamide is equally effective against the bacterial disease. The dosage recommended is 5½ gm/100 kg. Fish/day for a week.

c) Oxytetracycline: An antibiotic used in veterinary therapy is also commonly used for vibriosis and ulcer disease in fishes. The dosage recommended is 7½ gm/100 kg. Fish/day for one to two weeks.

d) Epsom salt: Magnesium sulphate when given mixed with diet is very useful in protozoan infection of the gut.

e) Di-n-butyl tin oxide also used as diet at a dosage of 25 gm/100 kg.

Fish/day for three days.

f) A homeopathic drug, *Argentum nitzicum* applied to the infected eyes of the fishes is found to cure the fungal infection.

g) Phenoxethol solution of 10 to 20 cc of 1% solution diluted in one litre of water is used to protect the fungal disease in fishes.

h) McGraw (1955) and Deufel (1963) have advised a daily dose of 10 gm. as sulfamerazine and 3gm. of sulphaguanidine for a period of 8 days in 100 kg. of trout.

i) A bath of Benzalkonium chloride (1 to 4ppm) may also be used for 1 hour to control the fungal infection.

Table 18.1: Common parasitic fish diseases, causative organisms and treatment

Causative organisms	**Disease**	**Symptoms**	**Treatments**
(1)	**(2)**	**(3)**	**(4)**
I) Bacteria	Fin and tail rot	White line on the margin of the fin; fin rays become brittle and start breaking.	1 minute dip treatment in 500 ppm copper sulphate solution.
	Ulcer disease	Open sores or ulcers on the body.	Dip Treatment for 1 minute in 1:2000 copper sulphate solution for 3-4 days.
	Dropsy	Accumulation of fluid inside the body cavity; scale protrusion; exophthalmic condition.	Dip treatment in 5ppm potassium permanganate solution for 2 minutes.
	Eye disease	Cornea of eye becomes vascularised and later become opaque; eye ball gets decayed.	Chloromycetin (8-10mg/litre bath for 1 hour for 2-3 days.
II) Fungi saprolegnia parasitica	Water-mold disease (Saprolegniasis)	Dies after ulceration or exfoliation of skin followed by haemorrhage; blindness; tufts of white hair like out-growth in the effected region	Dip treatment for 3 seconds in 1:10,000 solution of malachite green or for 5-10 minutes in 3% common salt solution or 1:1,000 solution

Causative organisms	Disease	Symptoms	Treatments
(1)	(2)	(3)	(4)
			of potassium permanganate.
III) Protozoans Ichthyophthirius multifiliis.	Icthyophthriasis (White-spot disease	Small whitish – cysts of about 1 mm diameter on the skin, gills and fins.	5 days bathing in 2 ppm methylene blue; hourly dip treatment in 1:5,000 formalin solution for 7-10 days.
Myxobolus lucioperace		Round or rice-shaped cyst on the gills	All infected fish should be destroyed or burnt.
M. exiguous	Knot disease	Small knots in the skin of common carp.	All infected fish should be destroyed or burnt.
Sporozoan	Pox disease	Small milky-white spots raised 1-2 mm above the surface of skin.	Intra peritoneal injection with 1ml. of a 1% solution of arsenic compound, Arycil.
Costia necatrix	Costiasis	Bluish-coating on the skin; lesions as irregular patches.	Bath in 3% common salt solution or in 1:2,500 formalin solution for 10 minutes.
M. pfeifferi	Boil disease	Large boils varying from the size of a nut to that of a hen's egg on several parts of body	Bath in 3% common salt solution or in 1:2,500 formalin solution for 10 minutes.
	Whirling disease	Caudal bend, deformity of the oral region and black ending of tail region.	Destroy all infected fish by applying quicklime (pond disinfection) at the rate of 2 t/ha
Myxosoma cerebralis		2 t/ha	

Causative organisms	Disease	Symptoms	Treatments
(1)	(2)	(3)	(4)
IV) Trematodes *Gyrodactylus* Sp. *Dactylogyrus* Sp.	Gyrodactylids Dactylogyrus	Fading of colours; drooping of scales; peeling of skin.	Dip treatment in 5% common salt solution or in 1:5000 formalin solution for 5 minutes
Diplostomum Sp *Diplostomum* Sp.	Diplopstomiasis (Black-spot disease)	Small black nodules of about 1-5mm diameter in the affected region.	Dip treatment in 3:100000 picric acid for 1 hour; Din-butyl tin oxide at the rate of 250 mg/kg fish.
Hemistomum spathaceum	Worm cataract	Whitish-eye due to the white dots, *i.e.* little worms situated between lens and lens capsule.	Dip treatment in potassium antimonyl tartrate (1.5 mg litre).
V) Cestodes Ligula sp.	-	Dull, sickly and with parts of alimentary canal swollen or completely choked by cestode cysts or worms (Fig. 97, H)	Dip treatment in 3:100000 picric acid for 1 hour; di-n-butyl tin oxide at rate of 250mg/kg fish.
VI) Nematodes Philometra Sp. Camallanus sp.	-	Dull, sickly and with parts of alimentary canal swollen or completely choked by round worms.	Dip treatment in 3:100000 picric acid for 1 hours; Di-n-butyl with oxide at the rate of 250 mg/kg/fish
VII) Acanthocephala Zelan echinorhynchus Sp.	-	Dull and sickly	Dip treatment in 3:100000 Picric acid for 1 hour; Di-n-butyl tin oxide attherate of 250 mg/kg/fish

Causative organisms	Disease	Symptoms	Treatments
(1)	(2)	(3)	(4)
VIII) Hirudinea Hemiclepsis Sp.	-------	Abnormal movements of the fish due to irritation as the parasites feed on the blood of host	Dip treatment in 1:1,000 solution of glacial acetic acid.
IX) Copepods Argulus Sp. Ergasilus Sp. Lereaea Sp. Caligus Sp. Pseudo cycnus Sp. Lernaenicus Sp. Clavellisa Sp.	-------	Loss of scales and presence of red spots; damage of gills.	Half an hour treatment in 500ppm formalin solution; mechanical removal by forceps followed by a bath in weak potassium permanganate solution for 2-3 minutes; bath in 1:1.,000 glacial acetic acid solution for 5 minutes and subsequent bath in 1% common salt solution for 1 hour; pond may be disinfected by applying Lindane at the rate of 8ml/1,000 litre.

Chapter 19

Fish Culture

INTRODUCTION

There is a large cultivable area of stagnant fresh water in this country in the form of ponds, lakes and tanks of which only small part is utilized for fish culture. It has been estimated that only 6,00,000 out of 16,00,000 hectares of water area is used at present.

Fish cultivation in India is expanding fast under the continues pressure of growing population. Traditional method are modernized and the farmers are trained to exploit vast resources of Inland waters fully.

Fish culture is the rational cultivation of fish in a confined water mass, where rearing is not left to mercy of nature. In its methodology fish culture can be compared with agriculture and animal husbandry. Fish culture involves the production of fish in a given water mass far above the level that is naturally obtaining. This is made possible by not only controlling the feeding growth and breeding of fish but also controlling the characteristics of the water mass. The success of fish culture as a trade lies mainly in the profit earned from the production. Today the importance of fish culture is being realized as never before. In South East Asia forages fish is being habitually eaten as the principal food along with vegetable and rice. Fish culture is practiced since ages. The importance of fish culture in the rural economy is being increasingly realized.

The importance of fish culture as a source of food production was then driven home more realistically and emphasis was laid on the need for extending fish culture activities to all the parts of the country with a view to develop the industry on scientific lines both in the private and public sectors.

India has a total potential for fish culture of about 3.6 million hectares of which fresh water ponds cover one million hec. reclaimable swamp 0.6 hec coastal and brackish water 2m. hec. The major areas of fish farming adopted may be classified as follows. (1) Fresh water pond culture (2) Paddy cum fish culture (3) Culture in sewage fed waters (4) Cage culture (5) Culture in reservoirs (6) Brackish water culture (7) Coastal marine culture (Mariculture) (8) Salt pan culture (9) Culture in cold water.

These practices have special feature of their own. Special technical knowhow is needed for only a few species that suit the requirements of each of the above methods. Most of the traditional methods in vogue until recent past were either of "stock and forget" type or based on faulty unscientific or empirical background. These are being corrected and improved upon on modern scientific lines. Thanks to the development of the infrastructure in the recent years.

ROLE OF FISH CULTURE

The Utilization of available Water Areas

A large number of fresh water ponds (Village ponds homestead ponds and irrigation Ponds), moats, mines, quarries, swamps, shallow areas along the canals and back water remain that merely serve as breeding grounds for mosquitoes and other disease producing vectors thereby becoming health hazards to the public.

At present only about 500 hectares in under fish culture. With a little more of investment and reclamation their waters can be profitable utilized for producing fish. Swampy areas can also be utilized for culturing the air breathing fishes.

The Utilization of Areas that remain as Waste Land

Water logged areas and flood plains which are not put to any use in agriculture, can form substantial, additional fish culture resources. Such areas which simply remain as waste land can be developed for producing carp and air breathing fishes by constructing suitable embankments and by guarding, the inlets and out lets. Shallow seasonal bodies of water which, get filled up during rains are common almost everywhere. Villagers are seen harvesting such areas for cat and air breathing fishes by cast nests, such bodies of water offer an immense scope for systematic culture, especially for air breathing fishes or when properly prepared for utilization as carp nurseries.

The Utilization of Organic Wastes

Human and animal husbandry wastes can be recycled for culturing fish and have two fold utility as manure and as fish food. Domestic sewage is directly utilized as fertilizer for fish ponds which give a high yield of fish. The combination of fish culture and livestock rearing ensure a high rate of protein food production economically.

Slaughter house wastes, offal, blood, kitchen wastes and sludge can form a good feed for fish. The CIFRI (Central Inland Fisheries Research Institute Barrack pore W. Bengal) has reported excellent fish production in ponds receiving these wastes without any manuring and artificial feeding.

The Utilization of Agriculture by Products

By products of agriculture, such as oil cakes of cotton seed, mustard, sunflower soyabean and Mahuva in addition to rice and wheat bran can be utilized as fish feed for enhancing the production of fish per unit area of water under fish culture. Horticulture is another field which can be integrated with fish culture. Pond banks can be utilized for growing, papaya, banana, guava and vegetables such as brinjals cauliflower and cabbage. The leaves of these plants can be fed to the grass carp. These crop will provide the farmers with additional income.

Source of Income of Farmers and Village Panchayats

Besides increasing the protein food supply in rural areas fish farming will create a new class of entrepreneurial occupation viz, fish farmers and thus it would create a gainful employment for those who are seasonally unemployed because of traditional agricultural farming.

By starting fish culture in village ponds, the panchayats can have additional income which can be used for constructing schools, hospitals and roads. The village panchayats can start fish farming in their ponds either by themselves of lease out the ponds for long period to individuals.

Individual farms can substantially raise their own economic standards through fish culture alone or through, its integration with other agro-animal industries and provide the villagers with jobs.

AIM OF FISH CULTURE

1. The main aim of fish culture is to obtain maximum yield of fish.
2. To obtain palatable and highly nutritive fish flesh.
3. By products of fishing industry.

CHARACTERISTICS OF FISH FOR CULTURE

The culturable fishes:

1. It should have a fast growth rate.
2. It should have ability to feed on natural food and artificial diet.
3. It should be hardy and resistant to disease.
4. It should be herbivorous in nature.
5. It should be able to tolerate a sudden change in climatic conditions and nature of water in ponds.
6. It should be fast growing and can attain good length and weight both.
7. It should be live in ponds with other fishes without any disturbance.
8. It should be prolific breeder and easing to breed in ponds.

9. It should be palatable and of much nutritive value.
10. It should not be predaceous.
11. It should be able to survive under temporary bad water conditions and hardships of transport.
12. Species that can be cultured together without competing with each other are preferable for maximum utilization of water.
13. It should support high population density in the pond.
14. It should successfully reproduce under confined conditions.
15. It should be able to withstand inter and intra specific competition.
16. It should have ability to live under changing physico chemical condition of water.
17. Assurance of supporting a dense population throughout life *i.e.* a social and gregarious fish.
18. Assurance of an easy and regular supply of fish seed *i.e.* the fish should reproduce in captivity or respond to induced breeding methods.

TYPES OF CULTIVABLE SPECIES

1. Indigenous or the native fresh water fish likes the carps.
2. A few salt water fishes that are acclimatized to fresh water like the chanos and mullets.
3. Exotic fish like the mirror carp, Chinese carp, Tilapia etc. are economically the most important fish for culture as:
4. They feed on zoo and phytoplankton, decaying weed, debris and other aquatic plants which are available in plenty in the ponds.
5. They are resistant to relatively high temperature and turbid water.
6. They can tolerate low oxygen content.
7. They have a fast rate of growth and attain a reasonably large size in the short time.
8. They breed profusely.

Important Cultivable Species of India

These species are called the major carps eq. *Labeo rohita*, *Catla catla*, *Cirrhina mrigal*, *Labeo*, *Calbasu*, *Labeo fimbriatus*, *Labeo bata* etc. Some exotic species *Cyprinus carpio Ctenopharyngodon idella*, *Hypophthalmichthys molitrix* (Chinese carp) *Osphronemus goramy* , *Tilapia mossambica* etc.

CLASSIFICATION OF FISH CULTURE

Fish culture practices may be classified into three different kind on economic and commercial consideration.

Extensive Fish Culture

Extensive fish culture is the least managed fish farming the pond may be large and little care is taken with regard to improvement. The modest yield raised on the natural food and then the natural production.

Intensive Fish Culture

Intensive fish culture on the other hand is one in which all attempt is made to achieve maximum production of fishes from a minimum quantity of water. It is best managed farm of fish farming in which fish is fed on artificial food and pond in improved by use of fertilizers. The production is very high and of high quality.

Semi-Intensive Fish Culture

Semi-Intensive fish culture is anything between the extensive and intensive fish culture.

CATEGORIES OF FISH CULTURE

The fish culture practices may be classified into six basic categories, depending on the nature of the water, mass used and the manner of farming.

Pond Fish Culture

In this case, the water mass consist of natural or artificially fresh water ponds. Fish ponds are of various types viz, nursery rearing production, stocking ponds, spawning ponds. The nursery and spawning ponds are shallow, others are moderately deep. The optimum size, depth and use of different ponds are given below:

Nursery Ponds

It is smaller and is used mainly to nurse the hatchlings for a period of two to three weeks until they become fry. The depth of the water column may be between 1.0 and 1.5 m. The maximum stocking density of hatchlings is about 10 million/hec. These ponds are meant only for a short-time. They could be used two or three times in a single breeding season. The nurseries can also be used as production ponds.

Rearing Ponds

In these ponds the advanced fry are raised for about 2 to 3 months. The ponds are deeper than the farmer type and have narrower sides to facilitate netting. As such ponds are located near the spawning and nursery ponds and their number may vary depending upon one, two or three year rotation of carp culture.

Production Ponds

These perennial ponds are the largest in size covering an area of over 91 m. and ranging in depth from 5 to 2 meter. Greater depth creates unfavourable conditions because at a depth of 3 to 4 m. The light penetration is poor and water warming is slower resulting in a poor production of food for developing fishes, such depths are however, permissible at the bottom of sluice.

Stocking Ponds

These ponds are similar to the production ponds in their areas and depths, but the area may vary depending upon the brood fish and the young fish required at the farm.

Culture in Fish Pond

The production of stocked fish mainly depends upon the suitability of species cultivated and the condition of pond where cultivation is done. It may be controlled and increased by way of maintenance and improvement of pond as well as by means of correct choice of fish, artificial feeding and rate of stocking. Three kinds of production are distinguished quantitative, qualitative and economic.

Quantitative production is characterized by a great quantity of fish as possible with no regard to the grade and quality of fish. All that is important is the total weight of the fish harvested.

Qualitative production has the main aim of production of fish of high quality *i.e.* of uniform size, weight and of high commercial value the quantity of fish is of secondary importance.

Economic production aims at producing as high a quantity of fish of a great market value as possible. Cultivable fishes should have the ability to feed on both natural and artificial diet and should be of nutritive value.

In pond culture, the fishes like major carps, chanos, mullets and some exotic fishes as mirror carps. Chinese carp, crucian carps are good cultivable fishes which suitable for fish water ponds. Much attention has been paid to the culture of *Labeo rohita. L. calbasu. Catla catla* and *Cirrhina mrigal.* Besides these other fishes like notopterus, chitala, labeo, mugil and corsula are also cultured. *Heteropneustes fossil's Clarias batrachus Anabas scandens* and *Channa punctatus* should not be cultured with carp in the same pond.

Culture in Man-Made Reservoirs

Manmade lakes provide about two million hectares of water and this would continue to increase. The water bodies after immense scope for fish culture. Large number of dams and reservoirs have been constructed during the recent years to provide water for irrigation and power generation. The successful fish farming is possible in dams and reservoirs. It is essential to make a detailed hydrological study of the water body. Suitable species that are stocked in dams are the major carps such as *Catla catla*, *Cirrhina*, *mrigal* and *Labeo calbasu*. These are capable of adjusting successfully to ecological conditions of the reservoir. The reservoirs in their initial stages are very rich in microorganisms which start dwindling after 5 or 6 years. It is therefore, necessary that new reservoirs are stocked on a large scale with fry of Gangetic carps.

In the absence of large stocking of major carps, carp minnows in the reservoirs are stocked. The major carp, *Labeo rohita* not been able to adjust to the environmental conditions of the reservoirs. But catla, mrigal and calbasu form a best assortment of species. Successful fish farming in reservoirs is necessary to impose restriction on size limit. Fish below a minimum size should not be caught. It has been estimated that by applying proper technique. It is possible to get an average yield of 75 kg/ha/year. Hence from about 2 million hectare of impounded water area available in this country, it is possible to get a yield of about 1.5 lakh tons of fish.

Paddy cum Fish Culture

In this case, the paddy fields that are inundated with water are used for cultivation of fish. The paddy cum fish culture is an ancient practice in the Far East, where extensive fields of paddy provide considerable potential for fish cultivation. Paddy fields fish culture practices are divided into three categories.

a) The paddy fields are used as temporary ponds after harvesting. Fish and paddy are not cultivated together but the fields used for paddy are later on used for fish culture.
b) The simplest method is to retain the wild stock that enter the fields during floods by erecting barriers across, the drains and channels. There is no regular stocking of fish in the fields. The fish that enters the field along with water are not allowed to escape.
c) The fishes are stocked in flooded fields and there is a simultaneous culture of paddy and fish in the same water.

The paddy fish practice is common in foreign countries and to some extent in India. Not all fishes are suitable for paddy cum fish culture in view of the special ecological and other conditions obtaining in the paddy fields. According to Coche 1967, the following criteria are generally used in selection of species for paddy cum fish culture.

1. Fish can tolerate high temperature and can support low dissolved oxygen content so characteristic of paddy fields specially those in low altitudes of tropical countries.
2. Fish can adapted to shallow waters necessary for paddy crops.

3. Fish has a fast growth rate so as to attain the marketable size in a short period.
4. Fish can tolerate high turbidity.
5. Fish can live in confinement of this nature and does not tend to escape.

Paddy fields to be used for fish culture are provided with strong barrier wall to prevent leakage of water and to convert the field into pond of desired depth. The field is also provided with trenches or ditches for the fish. These channels are connected with a small pond near the field to provide shelter to the fish against heat. Inlet and outlet of the fields are provided with screens.

The cultivation practices vary from place to place being subject to variation in-

1. Species of fish cultivated.
2. Variety of rice grown.
3. Choice between simultaneous or alternate cultivation of paddy and fish.
4. Stocking methods of the paddy fields.

The stocking methods of the paddy fields, are of two types:

a) In which wild fish are attracted to populate freely the paddy fields during floods and are then prevented to escape.
b) In which the paddy fields are stocked with fish seed or fingerlings either raised reared of collected from wild sources.

The suitability of fish depends upon their ability to withstand conditions of paddy fields. However, since the water in paddy fields is retained for a short time. The fishes selected for cultivation should have faster rate of growth. Species such as *Catla catla*, *Labeo rohita*. *Cirrhinus mrigal*, *Cyprinus carpio*, *Tilapia mossambicus*, Anabas, *Clarias batrachus*, and Channa species are widely cultured in rice field. The air-breathing fishes are considered good because of their accessory breathing organs, which helps them to breathe when the field gets dry and the water level in capture channel goes down. These fishes can survive well in shallow and turbid water of the rice field and give better cost of their raising.

Problems of Fish Culture in Paddy Field

One of the most common problem faced in paddy cum fish culture is the loss of fish during the period of their growth. The loss amounts to about 40-60% for the young's and 20-30% for the large fishes.

Fish yield is simultaneous culture with the rice is hardly profitable. The decrease in fish production may also be due to the fertilizers which are added to the main crop and preparations of the field being inadequately met.

Fish culture in rotation with the rice crop is however, a more efficient and profitable venture. It rules out various disadvantage of simultaneous fish culture and does not depend upon the rice crop for fertilization and intensification.

Brackish Water Fish Culture

Open or embanked estuaries with waters of varying salinities afford low lying, shallow fields under water for fish raising. Brackish water lakes are found along the coast line of India and are rich in fish fauna. The brackish water has certain advantages. It has rich supply of nutrient salts and there is little variation in pH. Brackish water pond is always alkaline its pH ranges from 7.5-8.3 with high salinity.

The water temperature of well irrigated ponds do not fluctuate greatly. The total area of brackish water in the country is estimated to be about 3000 square miles, and divided into two groups.

i) Open estuaries including the river mouths, lakes and back waters, which are always connected with the sea. They may perennial or seasonal.
ii) Embanked brackish waters which are confined areas of estuaries, lakes and back waters. Oxygen content of brackish water pond is dependent on temperature and salinity. The oxygen depletion is of very rare occurrence in these ponds.

The ecology of the brackish water limits fish culture to a few selective species. The most important factor is the factor of varying salinity whereas inflowing river water and rain water tend to dilute salinity the tidal water and evaporation increase the salinity. Temperature may be high or low, depending upon the depth of water and the climate the brackish water ponds also need, like the fresh water ponds, intensive management for high yield.

The general principals to be followed in the preparation of brackish water ponds for stocking are essentially the same as for fresh water ponds by regular draining and drying pond bottom the fertility can be greatly enhanced.

For successfully brackish water culture, selection of fish species is very important for which following important requirements are given below.

1. Availability of seed in sufficient quantities.
2. Compatibility with other species.
3. Hardness of the species to be cultured to tolerate fluctuations in environmental conditions like salinity and temperature.
4. Ability for rapid growth in brackish water.
5. Ability to utilize natural food of the environmental and can accept supplementary or artificial food.

6. High conversion efficiency ratio of food in to flesh.
7. Higher survival rate.
8. Market demand and valuable.

Brackish water fishes include the indigenous milk fish, Pearl, Spot, Bhokti, Tarpon and certain exotic fishes like, Tilapia, Hilsa, Pellona species, Mullets, Cat fishes and Perches etc.

In India about 9.47 lac hectares water area is in the shape of brackish water in maritime status. West Bengal has a highest of 23.28%, Gujrat 20.73% and Kerala 15.74%. However the fish productivity from the brackish water is not satisfactory, mainly because only 2.91% of total area is utilized for this purpose. There is also a scheme to develop brackish water. Fishing in some 9,02,000 hectares of water.

The brackish water of estuaries is highly productive. It also affords a suitable place for the culture of fishes for example, experimental culture of common carp, Silver carp, Big heads and Grass carps in the U.S.S.R., have given profitable results in numerous estuarine lagoons located especially on the southern side of the Black sea.

In India no attempt has been made for stocking of commercial carps. A further development based on principles of inland culture fisheries would make it possible to understand problems associated with the culture of commercial food fishes in brackish water.

In India brackish water pond known as bheris, ordinarily wooden sluice boxes with wooden shutters that can be manipulated by hand are used. It is a common practice in India to construct "V" or "W" shaped bamboo greeting behind. The sluice gates prevent the entry of undesirable fish or other animals.

Costal Fish Culture (Mariculture)

Culture of marine species of fish in marine coastal water in its infancy and has been initiated as pilot projects. These are great potential for culture of fish, prawns, pearl, oysters and mussels. Such water are suitable for mariculture are especially abundant in Kerala (Pokkali fields). Goa (Khazanlands), Karnataka (Kharlands) and West Bengal.

Marine fishery resources are self-renewing and can be harvested year after year without depletion. On the other hand agriculture depends upon two diminishing and non-renewable resources, viz. the land and the fossil fuel. Hence, we have to exploit our sea resources which are quite extensive. Sea contains billions of phytoplankton which entrap solar energy by photosynthesis and lead the production of zooplankton. The phyto and zooplankton from the food of fishes. Crustaceans and other animals which can be utilized by man.

Marine culture is a recent practice, and still in its infancy. Japan has brought a revolution with commercial culture of pelagic yellow tail and other notable marine, culture that include Salmon, trout and cod in Norway. The salinity, pH, temperature thermadyne and up welling are the important factors influences fish production of coastal area. Salinity affects osmoregulation of fish. Some fish are euryhaline and can withstand wide ranges of salinity but in general there is optimum salinity for fish along the coastal areas. Salinity affects the buoyancy of the eggs.

Generally the pH of sea water is 7.5-8.4 along the west coast of India, the pH is 8-8.3 which is favourable for fish.

The temperature usually ranges between 23°-29°C touching a low 21°0 C during winter.

Thermocline is the zone of water layers separating in a vertical gradient, the upper zone of warmer water distinctly colder water the deeper region of the sea.

The upwelled water has a low temperature, higher density, low oxygen content, and is rich in nutrients, consequently, higher biological productivity and higher fisheries along the west coast of India, is the result of upwelling of water. Upwelling is reported to occur in east coast of India but nutrient rich water is not present.

The mariculture experimental achievements have paved the way for commercial culture of *Sardinella longiceps*, *Sillago sihama*, *anguilla bicolor bicolor*, *Chanos chanos* and *Mugil cephalus*.

In pollution free water near shore, coastal and open sea environments, especially of India offer wide scope for adopting fish and shell fish culture operations. Floating cages for the culture of fishes, prawns and lobsters, racks for oysters, raft rope pole and long lines for mussels, nets for sea weeds are at present widely employed.

Salt Pan Culture

In off season of salt manufacture the salt pans are utilized for fish culture. The main categories of culture for *Sillago*, Salt pan culture for *Chanos* and Mugil running water culture for Anguilla are in vogue.

Salt pans, as those at Neel larvae and Bheemunipatnam (A.P.) are especially useful for culture of prawns and fishes on a commercial basis. The culture way be undertaken during the period when the manufacture of salt is suspended. The off season is salt manufacture lies generally between June and December. During this six-months period, an yield of 100 kg/hec. of *Chanos* and *Mugil* has been experimentally achieved.

SPECIAL OPERATIONAL TECHNIQUES OF FISH CULTURE

Cage Culture

The fish to be cultured are held in a section of water of a flowing river by being imprisoned in a cage of metal or bamboo structure, supporting wire, nylon or split bamboo mesh mats. It is rather recent practice but with great promise cage culture has been practised in Cambodia rather traditionally. Cage culture has opened new ways of easy exploitation of natural waters both fresh water and salt water which are more suitable for fish raising than the waters of artificial ponds or the like. The one very great advantage offered by cage culture is the great density of stocking that is impossible to maintain in pond culture.

It is the running nature of water which is the key to greater production. The running water keeps a constant supply of food, a regular removal of wastes and a quick replenishment of what is used up by the caged crowded fish. Thus serious problems arising out of population densities common in pond culture do not simply exist in cage culture management and therefore, investment of capital is minimum. The greater yield obtained from cage culture in progressed by the following characteristics:

i) High densities of stocking of fish.
ii) Effective water area of volume available to fish being much larger than the actual area or volume enclosed in the cage.
iii) Relatively simple and cheap technique application to both fresh and salt water.

Cage culture has also made possible culture by pond culture method.

Cages are made of galvanized steel wire frames with nylon meshes or simply of bamboo frames with split bamboo mats. The cages are either fixed to stake or are left floating by means of strong floats.

The size of the cage varies greatly. The commonly used size has a surface area of about 10 sq. meters.

Cage culture is being practiced for a variety of fishes:

a) **Carps**: common carp (Cyprinus carpio) in Indonesia Japan and Russia, Burma spp. In Cambodia.
b) **Cat fishes**: Pangasius in Cambodia and Thailand, *Clarias batrachus* in Cambodia, Channel fish in U.S.A.
c) **Clupeids**: Yellow tail (*Seriola quinqueradiata*) Trout (*Salmo* spp.) and pacific salmon (*Oncorhynchus* Spp.) in Japan.
d) **Other's**: Anabas, Ophiocephalus and Oxyeleotris (goby) in Cambodia, Tilapia in Egypt and Guatemala.

Cage Culture Method Adopted in India

Cage culture has proved a boon for reclaiming extensive waste water *i.e.* swamps and derelict water of Assam, Bihar and Karnataka about 0.6 million hec. Channa and Anabas are responding well to cage culture practices. Cages are made of split bamboo mat supported on frames measuring 2mx1mx1m. The cages permit easy harvesting of reared fishes. Cage culture of cat fishes (Mystus) is also being explored in rivers.

Cage Culture has certain Advantages

1. The enclosed fishes can be prevented from being taken away by predators.
2. They can stocked at high density rate.
3. Water passing through cages brings food supply to fish white it carries away the wastes.
4. Management and capital investment is less.

Pen Culture

This is simply an enclosure often along banks and shores of some device which utilize the bank or the shore line as one of its sides. Fishes are then raised in this enclosure.

Pen culture has become popular is some of the Asiatic countries. The traditional polyculture concept is pen culture system. High production in pen culture is related to dissolved oxygen and exchange of water. As in open water, oxygen in hot a problem, the production can be increased in pen culture.

The procedure used in pens are bamboo poles fixed in circle or rectangularly covered with nets. The sinkers are to be put into 20 cm. depth of silt and sometimes anchors are placed with a distance of 0.5 m for reinforcing the pen. The bamboo poles should be about 0.6 m. below the silt soil bottom. In order to prevent heavy waves in open water pens, some aquatic plants is placed at least 10 m away from the pen. In common practice. Chinese use two layers of net enclosures of some mesh size. The inner one is used for culture of fish. The ratio between cage and pen is 3:11. The average net pen yield is 4170 kg/hc/yr. (Lisifa 1990). The production can be further increased through flow rate of water in pen management.

NUMBER OF SPECIES OF FISH CULTURE

There is another way of classification based on the number of species cultivated of a time. These are three types

Monoculture

The production of fishes for commercial purpose, may be either to produce all males or all females, depending upon which sex has better food conversioners. Monoculture is the cultivation of only a single type of fish species in a pond it may be Tilapia of one species, common carp or any other species. In monoculture fishes can be all of different ages and life stages or they can be separated in different ponds.

The monoculture can be done by the following method:

i) The sex of the fish is identified before maturity, and male and female are separated. The faster growing sex can be reared in tanks and reservoirs.
ii) Other method is to feed they fry on a minute amount of sex hormones so as to produce all males of all females. This is a reliable method.

Correct does is used at the proper time and for the required period. The gonad of a teleost is an undifferentiated one as in other vertebrates and this undifferentiated gonad can be triggered to form ovary or testis by hormonal treatment.

An indirect method is also used by fish farms to produce female. In trout, genetically female fish are converted to functional males by feeding methyl testosterone at a dose of 3 mg/kg food. The milt produced by these males is used to fertilize normal eggs. As Y Chromosomes are not present in the sperms only females are produced by the fertilized eggs.

If sterile fish are required, the eggs of these females are subjected to heat stock so that all female triploid fish are produced, in which gonad development does not take place. The monoculture fishes includes *Carassius auratus*, *Oryzias latipes*, Tilapia, Salmon, Trout etc.

Monosex Culture

Monosex culture is the culture of only one sex of a species of fish in a pond. When only the male or female individuals are stocked together all the energy of fish goes into growth and not in reproduction. Tilapias are often used four monosex culture because they attain maturity at an early stage when they are small in size. It they are separated by sex, they continue to grow and attain bigger size for use as food or for the marketing purpose.

Most of the investigation have used hormones to induce egg and sperm production in fish, but the precocious sexual maturation and the inhibition of gonadal development are also important for aquaculture.

If gonadal development is prevented, all food will be converted into flesh and there will be on wastage of food by its conversion into gonadal material. Hence, production of a sexual or sterile fish would be an advantage in commercial fisheries. Surgical gonadectomy is of no use for commercial purpose. Certain steroids as estradiol **1-17-B** in high does and methyltestosterone inhibit spermatogenesis in the Trout, Salmo gairdneri. But the use of exogenous steroids to inhabit gametogenesis is not favored for commercial production as these not favored for human consumption due to undesirable chemical residue in the fish muscles. The inhibition of gonadal maturation is the use of non-hormonal non-steroid chemicals as methallibure and cyproterone acetate. Both these chemosterilants reduce gonadotropin production and lower androgen secretion, but do not prevent gonad maturation.

Poly Culture

In polyculture two or more species of fishes are cultivated together in a pond. Stocking of fishes of different species which differ in feeding habits in same pond is called the polyculture pond. It is based on the principle that when several species belonging to different ecological niches of a pond are cultivated together, the available food in the pond is utilized most efficiently leading to maximum production. The stocking is done in such a way that towards the end of vegetative period, marketable fishes of all the species are obtained. In all the instances of composite fish culture and particularly in the additional fish polyculture careful observation is needed throughout the period of their vegetative growth. The factors suggested in foregoing account must be taken seriously to prevent disappointment in a rather new technique of polyculture. The example, the raising carps with other fishes increases fish yield significantly when compared to monoculture of carp. A wider introduction of polyculture in practice of pond fisheries at fish farms would make it possible not only to increase the quantity of pond fishes and productivity but would also ensure higher income with lower investments.

In India, polyculture is an old practice, and species like *Catla catla* (surface feeder) *Labeo rohita* (column feeder) and *Cirrhina mrigal* or *Labeo calbasu* (bottom feeder are generally stocked together in the same pond).

These fishes, when stocked in a ratio of 3:3:4 utilize food from all the levels of the pond. At the same time they do not harm each other, nor do they complete with each other for food. On the contrary if a single species is stocked, the food is consumed only from a restricted level of the pond to which the fish belongs. Among the other fish species which can be reared with the carp the preferred ones are the predatory cat fishes and murrels. As the letter fishes consume trash fish they are introduced only after the fingerlings of carps have grown sufficiently larger in size. So, they may not fall victim to predatory fishes. Supplementary feed in the form of larvae of insects, chopped flesh and worms etc. may be provided to feed these fishes.

CLIMATIC FISH CULTURE

Fish culture practice may also be classified on considerations of climatic factors.

Cold Water Culture

This involves culture practices conducted in springs pools or lakes of hills of high altitudes. Cold water fisheries have a great scope in the upland water masses of India, which are significantly abundant in the Himalayan region of Kashmir, Himachal Pradesh, Uttar Pradesh, North Bengal, Sikkim, Bhutan and Arunachal Pradesh. These water bodies have cold water that can sustain only the so called cold water fishes.

The Indian major carps do not survive there due to low water temperature. The water bodies of the hills are characterized by low temperature, high oxygen content low carbon dioxide and high velocity. The rivers and streams, of the hills are shallow, with clear flowing water and rocky bottom. Thus, ecological conditions of the hills streams are entirely different than those of the plains. Fishes commonly occurring in the hilly waters and those found in the hills have adapted themselves for the peculiar condition of the region. A large number of species occurs naturally in the cold water for example, Trouts, Mirror carp, *Tinca tinca*, *Carassius carassius*, Mahsheer, Snow trout, Barbilus, Labeo, Garras, *Glyptothorax pectinopterus*, Nemachelius, *Puntius ticto Channa gachua* etc.

Warm Water Fish Culture

Both lakes and streams have not springs such thermal water are widely distributed over the earth and are warm, clear, highly mineralized and generally devoid of oxygen. Fishes generally are not known to be established in water warmer than 104°F, although they may survive short spells in water upto 110°F. Although reports of fishes exist for water from 110°F to 164°F. Close examination has failed to support any of the claims. Streams are an immense source of fresh water fish production. The species of fishes that include the major carps, murrels carps, murrels, cat fishes and Osphronemus etc. can migrate any place suitable condition.

OTHER TYPE OF FISH CULTURE

Running Water Fish Culture

Running water race ways or closed circulated system are highly productive being practiced in many parts of Asia. Three different types of running water fish culture system. Such as open type, semi closed and closed circulated system are prevalent in the Asiatic region.

Structural component in running water system

a) The main structural component of closed running water system it consist of
   i) Running water cemented rectangular tanks.
   ii) Earthen pond area.
   iii) Aquatic weed cultivated water tank.
   iv) Filtered storage water tanks.
   v) Pump house for pumping water.
b) Open type of running water systems are usually constructed near the river site. Various species of fishes such as carps, cat fishes, tilapia, eels, sea bass and sea bream are taken. In China, the main species taken in running water system are grass carp and Wuchang fish. Polyculture along with fresh water mollusc culture were taken in earthen ponds placed adjacent to the running water tanks. The stocking density is usually 5-10 kg/m2 of area and the stocking size of fish is 50 gms. onwards. There tanks are also used for fingerlings rearing in some parts of Asiatic countries.

In closed running water system with grass carp and Wuchang fish, the maximum production so far obtained in China 45 kg/m2.

However, with *Tilapia mossambica* the yield is 103 kg/sqm per year in Poyang city in China. In running water system, It tidal water or flooded water comes to pond in a controlled rate of 2000 cubic meter of water per day the production can be raised more than 49 tons/he/yr and if tidal water flow in to pond is reduced to 200 kg/m^3 the production is reduced to 26 tons/he/yr. Being suggested by Pearl River Research Institute, China.

Air Breathing Fish Culture

Derelict shallow waters with poor oxygen content could be utilized by culturing air-breathing fish species such as *Channa marulius* and *C. striatus*. *Clarias batrachus* and *Heteropneustes fossils*.

Culture of air breathing fishes is only occasionally practiced in India. The air-breathing fishes such as Murrels Magur, Singhi and Koi form an economically important group of fishes. These are marked alive and are therefore called live fishes. They breathe gills and accessory respiratory organs. On the contrary these fishes are generally eliminated from the pond because of their carnivorous diet. These fishes are cultured in fish farms that would prove a more profitable venture for the following reasons

a) They have a good market demand.
b) They are more nutritious and of high medicinal value.

c) As most of them have air-breathing organs, their culture can be practiced in rejected foul water of shallow ponds, swamps and in other kinds of oxygen depleted waters.
d) They are highly recuperative.
e) They grow faster and require little rearing.
f) Culture involves low risk and simple management.
g) When cage culture practice is adopted, the income may increase manifold.

The cultivable air breathing fishes includes various cat fishes and the murrels. These are Clarias batrachus and Heteropneustes fossilis among the cat fishes and various species of Chana among the murrels constitute an economically important fisheries.

These fishes being provided with air breathing organs may be cultivated in marshy and swampy places which constitute about 0.6 million hectares of unutilized water of the country. Reclamation of these water potential is possible also by cage culture method of raising air breathing fishes.

ADVANTAGE FISH CULTURE

1. Fish is a poikilothermic or cold blooded animal so need not spend any energy for temperature regulation. It does not spend energy to maintain its position in water column. The specific gravity as fish is nearly same with that or water. But land animals are subjected to walking resistance and gravitational force.
2. There is security as catching fish from the cultured water, unlike the uncertainty of catching in natural vast waters. That too, culture of fishes is more rapidly expanding in interior land based resources of the country away from all coastal belt.
3. Unlike major agricultural carps, fish do not consume water, compared to any bird and mammals used for husbandry purposes. Fish have the highest fecundity.
4. Edible tissues of fish is appreciably greater than that in mutton, beef of poultry. The lean meat percentage in fish (80.9%) is greater than chicken broiler (64.7%) and Beef (Choice grade) 51%. The refuse loss from fish (13.7%) is lesser Beef (15%) and chicken broiler (32%).
5. Fish can convert food into body tissues more efficiently than other farm animals. The reason for the superior feed conversion efficiency of fish is that it assimilates diets with higher percentage as protein apparently because of their lower dietary energy requirement. These requirements for metabolism are less in fish. Which evolve in a protein rich energy deficient environment than in warm blooded animals, because fish do not have to maintain

constant body temperature. Fish excrete relatively little energy to maintain position in water.

6. Fish excrete nitrogenous waters as ammonia instead of urea or uric acid that is less energy expenditure in protein metabolism. Because, of their low energy requirement. Fish can channelize more energy to wards anabolism that is fish can synthesis more protein per calorie of energy consumed that poultry or livestock. Thus the primary advantage of fish over land animal is low energy cost as protein gain and superior feed conversion efficiency.

Chapter 20

Chemical Constituents of Fish

INTRODUCTION

The nutritive food value of fishes have been excellent sources of protein for human diet. This protein is relatively of high digestibility biological and growth promoting value for human consumption, nutritional studies have proved that fish proteins rank in the same class as chicken proteins and are superior to milk, beef protein and egg albumen. Fish protein having the essential amino acids in desirable strength for human consumption namely lysine, arginine, histidine, leucine, isoleucine, valine, theonine, methionine, phenylamine and tryptophan. This accounts for the high biological value of fish flesh. The fish flesh is a valuable supplement to human diet for people who are habitually taking cereals and sugar, the fish flesh having some minerals, Iodine Vitamins and fat. Over and above all, fish flesh cooks easily and offers portable taste and flavour and easily digestible. In fresh fish and preserved fish there is a difference in nutritional value. Flavours and biochemical composition, do not remain normal and changes during preservation processes and storage. Then some species of fish may show variation in biochemical composition of its flesh. They constitute important sources of oils of medicinal and industrial importance. In America, large quantities of fish are caught only for the oil and meat. Comparatively very small quantities of fishes are utilized for non-food purposes in India.

PRINCIPAL BIO-CHEMICAL CONTENTS OF FISH FLESH

The principal biochemical contents of fish flesh are protein, fat, and water. The flesh having protein content 20% fat and watery vary widely and one varies inversely to the other. There are minerals, vitamins and enzymes, carbohydrates does not occur in the flesh except as glycogen in the liver. Water content varies from 55 to 83 percent. This is relatively high as compared to that of birds and mammals.

FISH FLESH CONTAINS

It contains protein 15-25%, moisture oil 80% minerals 1-2%, and other constituents 1%. The water content of the fish is known to very inversely as the fat content, other constituents do not vary widely. Vitamins and enzymes are also present.

Proteins

95% of the protein nitrogen of fish muscle can be extracted by Sodium Chloride Solution. The intracellular protein consists of.

Myosin (70%), myogen (6%), myo albumin (7%), and globulin (8%) collagen and elastin (3%) (connective tissue fibres) constituting the stroma proteins. Other proteins *e.g.* phospho protein, nucleo proteins and hemoglobins are present in minute quantities. These proteins are either simple or conjugated. Latter is characterized by a substance to which protein molecule is attached to nucleo-protein has chromatin as a substance.

The simple proteins are (albumen) soluble in water or salt solution, acid and alkalies (globulins). The fish myosin which has the same mobility as rabbit myosin has been fractioned into 'acting' and 'true' myosin.

Atleast two tropomyosins have separated. The digestibility coefficient and biological value of protein varies from species to species, for example pelagic species (Sardine, Herring, Mackerel and Tuna etc) are characterized by the presence in the large concentration of basic amino acids particularly histidine which is largely responsible for the meaty flavour of fish flesh. The protein nitrogen amounts to only 2 to 3% of muscle weight. Pelagic species are characterized by the presence of basic amino acids which give characteristic meaty flavour. During spoilage, histidine, under goes de carboxylation to histamine, which is probably responsible for food poisoning.

The protein is contained mainly in the skeletal muscles in fish flesh, which is of two types. Red muscle and white muscle, which differ in the presence of myoglobin, The red muscles has glycine, leucine, organine and phenylamine, whereas white muscle has more lysine, aspartic acid and glutamic acid. The protein content of the muscle extractable by soluble fractionation is generally divided into 3 groups.

1. Albumin fraction comprising 16 to 22% present in sarcoplasm and interstitial fluid, extractable with weak salt solution.
2. Protein fraction comprising 75% present in the contractive elements, extractable with strong electrolyte solutions.
3. Stroma comprising 3% present in myocommata cell membrane and connective tissue extractable with difficulty.

The above three kinds of muscle proteins are not pure but mixtures.

The Albumin is shown to consist of six components one protein fraction of the contractile elements is shown to consist of myosin actin, actomyosin and tropomyosin fractions. The myosin is obtained as a supernatant part.

The actin is obtained by the acetone method, the effect of ATP on actomyosin is supposed to be an important substance for fish spoilage. The tropomyosin is extracted by dehydration of organic salts. The protein of a few species of fish has been found to be rich in lysine and methionine. The fish proteins are superior to egg albumen, beef protein and chicken proteins.

Non protein nitrogen forms 7.5 to 17.1% of the total nitrogen. The pattern of distribution of free amino-acids in the non-protein nitrogen fractions varies with particular fish species. Aqueous extracts of fish muscle have been included in different nitrogenous constituents *e.g.* volatile bases (amino-mono-di-and trimethylamine), Trimethyl ammonium bases (trimethylamine oxide and betaine) glyoxaline derivatives (Carnosine, anserine and histidine), guanidine derivatives (creatine methyl guanidine etc), urea, amino acids and purine derivatives.

Planktophagus fish have higher protein contents than other fishes with other feeding habits, lean fishes have greater protein proportion than do fat fish. The cultivated fishes have higher protein contents.

The number of non-protein nitrogenous components include:

a) Peptides to which fish owes for much of its flavour.
b) Free amino acids of which histidine is important. Red muscle is richer in histidine than white one.
c) Volatile bases like ammonia, trimethylamine particularly in marine fishes.
d) Creative, taurine, betaine, histamine, glycine, uric acid carnosine, anserine etc.
e) Nucleic acids.

Creatine phosphoric acid is present in aqueous extracts of many fishes, fish with pronounced migratory habits usually contain higher, concentration of the same than other. Alpha amino butanoic has been identified in the alcoholic extract of sun-dried Bombay duck (*Harpodon nehereus*).

The nitrogenous constituents of fish muscle are qualitatively similar to found in the muscle of higher vertebrates but differ in containing more collagen and less extractives. The distribution of nitrogen in different species of fishes appears to be similar. The alpha amino nitrogen varies from 65% to 75% of the total nitrogen.

Carbohydrates

Fish flesh contents, small quantities of carbohydrates. Glycogen is present in living fish and rapidly changes to lactic acid after death.

Enzyme

Fish muscle contains appreciable quantities of proteolytic enzymes.

Amylases, Lipases, Thiaminase adenosine triphosphatase, choline esterase and other esterases, nuclease butyrinase and glycogenase are also present thiaminase occurs in fresh water fishes but absent in marine fishes.

Mineral Constituents

The mineral content of fish flesh varies from 1 to 2%. The main constituents are calcium, Phosphorus, Magnesium, Sodium Potassium, Sulphur and Chlorine Traces of other elements like molybdenum, Manganese, Zinc, Aluminium, Barium, Silver, Silicon Mercury, Cobalt Vanadium, Nickel, Iodine, Lead Radium, Titanium, Bismuth are also found. Cadmium and Gold are present in some species. Boron, Fluorine, Bromine, Lithium, Strontium are also present in marine fishes, than in fresh water fishes.

The mineral constituents calculated on the bases of 100 gm. of protein in the fish flesh, are Calcium 0.109, Potassium 1.671, Magnesium 0.133, Phosphorus 1.14, Sulphur 1.119, and Iron 0.0055, Iron, Calcium and Phosphorus are present in readily available forms; 3.4% of the available iron occurs as an iron-nucleo protein complex.

In the bones and scales of teleost fishes have calcium in the form of apatite while in otoliths. It exists as carbonate. The Phosphorus occurs in fish as phospho proteins, phospholipids. Complex phosphoric acid (as Vitamin B_1 & B_2), adenosine-polyphosphate and glycerol phosphatides. Other mineral constituents are also present in fish muscle. In fresh muscles copper content ranges from 1.4 to 4.1 mg. (av. 2.5 Mg/Kg), Iodine from 130 to 5130 μg/kg. Fluorine 5 ppm, Arsenic 7 ppm occur in organic combinations and do not give rise toxic effects. The flesh of some Indian fishes contains 0.0382 to 0.0502 mg/100 gms. of Manganese, 34-66 mg/100 gm. of sodium and 119.231 mg/100 gm. of potassium.

Pigments

Xanthophyll, pigment present in fishes are lutein in the form of as taxanthin and a taraxanthin like pigment. The red pigment in salmon has been identified as a form of as tacin (4,5 4,5-Tetra Keto-B-Carotene, $C_{40}H_{48}O_4$ m.P.243). Marine and fresh water fishes differ little in their carotenoid distribution.

Vitamins

Fish is a good source of vitamins A, D and B complex all are essential for human diet. Liver is rich in vitamin B_{12} and B-Complex, Vitamins A and D. Vitamin B-Complex includes thiamin, ribo-flavin and nicotinic acid. The distribution of vitamins is not uniform in the body of fishes. Vitamins A and D occur in inner organs vitamin B occurs in liver, eyes, skin, kidney, spleen, pyloric caeca and intestine. The vitamin content varies in species. Example eels are rich in Vitamin A but mackerel in vitamin D marine fishes of India are rich in vitamin B_1, B_2 and D.

The flesh of marine fishes bears thiamine, riboflavin and nicotinic acid for. Example Indian Shad, Pomfret. The choline, Vitamin B_{12} present in *C-batrachus* (Indian fresh water fish), choline is known for medicinal value. Certain fishes like tuna, dog fish, Pollock, cod are rich in vitamin A_1 and A_2.

Phospholipids

The phospholipids content of the flesh present in various species of fish vary from 0.17 to 0.7% on the flesh of cod reveal that complex mixture of phospholipids is present in which lecithins, predominate (50-60% of the total phospholipids).

A group of unidentified phospholipids designated as phospholipids A-B and C are separated, A and C may be the derivatives of Bio – phosphatidic acid. Three types of inositol phosphatides have been identified. Phosphatidyl ethanolamine is a minor component and plasmalogens are present in traces. The fatty acids of the lecithins are remarkable for their high content of C_{20} and C_{22} polyethylenic acids and for the absence of hexadecanoic acid. Phospholipids A contain high proportion of stearic acid, while B is virtually devoid of C_{22} acids. The fatty acids of phospholipid C resemble those of the lecithins.

The phospholipids found in the flesh of 22 species of Bombay fish are as follows. (dry basis). Total phospholipids 4.3-5.7 lecithins 2.3-3.6, sphingomyelin, 0.2-0.6, and cephalin 0.9-2.2%. Total cholesterol in the flesh varies from 0.3 to 0.8% and free cholesterol 0.15 to 0.50%.

Fats

The fat content present in fish is distributed in tissues. It is present in large quantities.

Fish Oils

They are triglyceride, esters of fatty acids combined with sterols, hydrocarbon, vitamin, phospholipids and colour substances. The oil content of marine and fresh water fish differ in relative proportion of various fatty acids. Marine fish oil contains, larger quantities of C^{18} C_{20} and C_{22} acids.

Fresh water fish oil, contains large quantities of palmitic acid and stearic acid but a smaller amount of C_{20} and C_{22} acids. The fish oil differ from vegetable oils in containing a wider variety of fatty acids particularly of the highly saturated groups. Shark oil is rich in unsaponifiables substances, whereas sardine and herring oils are poor in this.

CONCLUSION

Fish contents helps in making it of great importance due to its medicinal value and nutritive food value etc. that is why they are consumed in India and abroad.

Chapter 21

Exotic and Transplanted Fishes

INTRODUCTION

Several fish species have been imported and introduced into the water of our country. This culture is known as "Exotic fish culture". Several exotic species have been introduced successfully into different parts of India. Exotic species are mostly superior quality of fishes and they can tolerate climatic changes and quickly adapt to the new environment. Indigenous species of fish are definitely economical and easy to culture. Still there is need to culture exotic fishes. Transplantation of certain species of fish have also been practised in India and it has been found to be quite profitable. Labeo has been successfully transplanted from Bengal to Chennai. The fingerlings of milk fish *Chanos lates calcarifer* (Bakti), *Mugil cephalus* and *Polynemus tetradactylus* have been acclimatized to fresh water. The milk fish and the Bakti have been found to grow well in fresh water.

However, great care must be exercised in introducing exotic species for culture. Since they might disturb the biological balance of the existing species. For example introducing of Tilapia into carp culture areas in India adversely affected the survival of mrigal (Cirrhina) when the two species were stocked together. Hence, a detailed study must be made.

Regarding the overall suitability of the exotic species before they are introduced into an area. The exotic fishes also include the game fishes (*Salmo trutta*, *Onchorhynchus nerka* etc), food fishes. (*Tilapia mossambica*, *Ctenopharyngodon idella*, *Cyprinus carpio*, *Puntius javanicus* etc) and larvivorous fishes (*Gambusia affinis*, *Lebistes reticulates* etc.).

IMPORTANT EXOTIC FISHES

Trout

Trouts are the indigenous fishes living in the mountain waters Contrary to this opinion the cultivation of trout is no longer tied to the mountains only. In recent years salmonid farms particularly the Danish and the French salmon farms are enjoying a high reputation by providing the largest production and the best return per acre of their cultivation.

There are three main species of salmon which are important game fishes introduced in India. These are *Salmo gairdneri* (Rainbow trout); *Salmo trutta fario* (Brown trout) and *Oncorhynchus nerka, Salvelinus fontinalis* (Brook trout).

S. trutta fario was introduced by the year 1863 and in Kashmir (1901) and Kerala (1909) from England. But they were successfully transplanted in Nilgiris in 1960. There are six reservoirs of trout in Nilgiris.

Salmo gairdneri have been imported from various places like Ceylon and Germany in 1968 and they are transplanted successfully in Nilgiris.

S. irideus were also imported from Europe and introduced into Kashmir in 1940. *S. salar* was transported from N. America 1968 to Kashmir and *Oncorhynchus nerka* from Japan (1968) to Nilgiris.

Salvelinus fontinalis has also been introduced into Kashmir and Himachal Pradesh. Ova and fingerlings of trout were introduced in Bhowali, Nainital, Neokochia tal, and sat-tal of Kumaus but did not prove to be a success. There are several trout hatcheries in India as in Nilgiri, Kerala, Kashmir and Himachal Pradesh. The intensive culture of trout is now being carried out throughout the world.

Carassius carassius

This is commonly called the crucian carp or golden carp and grows well in the cold water. They are also introduced into the plains but reported to have breed at Sunksuta fish farm in Andhra Pradesh. It was introduced into the Ooty lake in a Chennai in 1874 from Central Europe. It is well established in Nilgiri waters. These crucian carp survive in water which is poor in oxygen contents and grows slowly. Attempts made to introduce the fish is plain water met with little success. The flesh of fish is scarcely appreciated. They are able to breed throughout the year.

Tinca tinca

They are known as tench or Doctor fish and was first introduced in 1874 in the Ooctacamund (Chennai) from England. They have now fully acclimatized into cold waters of Nilgiris, but is reported to be breeding and living in the sunkesula. Fish farm Andhra Pradesh. These fishes are bottom feeder and consume insects, vegetables, debris and planktons. They grow slowly and can be cultured with other suitable carps the grow much more slowly than the carps and attain a maximum length of 50 mm. They require a temperature of 20°C or even more to reproduce and so they breed quite late in the months from February to April.

Cyprinus species

This is also called the common carp and is the native of temperate regions of Asia but it is now conveniently cultivated in different parts of the world. There are three varieties of this fish.

a) Cyprinus carpio (Mirror carp)
b) Cyprinus carpio (Scale carp)
c) Cyprinus (Leather carp).

The mirror carp was first introduced into the Nilgiris in 1939, from Ceylon and stocked in Ootacamund lake. Later on they were carried to Bangalore and in 1947, theywere introduced into Nainital lake and other lakes Kumaun Himalayas. The mirror carp is nowwell established in Himachal Pradesh, Punjab, Jammu and Kashmir, Madhya Pradesh, Rajasthan, Delhi, Bihar, Manipur and Sikkim. It is an ideal fish for introduction into hilly regions, being an omnivorous, non predator and fast growing species. It breeds in confined waters. The scale carp was brought from Bangkok in 1957 and stocked in Cuttack. This variety is fast growing and omnivorous. The leather carp is also successfully cultured in ootacamund lakes.

These species can be cultivated in monoculture or polyculture. They may be cultivated in ponds, cages, paddy fields, sewage fed water, streams and with the duck in poinds.

Intensive culture practise in India yield 1500 kg. per hectare, in China 7500 kg per hectare and in Indonesia and Japan, upto the mark of 4,000000 in species types of culture practices like as incages and in sewage fed water. The common carps generally breeds twice in a year. In India the first season varies from July to August and second from mid, January to March. The eggs are released on plant roots hanging just below the surface of the water. Hatching occurs in about 2 to 3 days and the yolk sac is absorbed in about 6 days after fertilization. Fry feed on the zooplankton and later on any kind of food available in the pond. Common carps are probably the easiest fishes to keep breed and harvest. Mature fish continues to breed until it is too old to spawn. It feeds on all kinds of food ranging from zooplanktons to decaying plants and grows to length of 0.9m. under the intensive culture practice.

Hypothalmichthys molitrix

This fish is also known as silver carp. It is a food fish grows rapidly and valued for its flesh. The silver carp is a native of china, but has now been introduced into Burma, Philippines, Hongkong, Japan, Malaysia, Pakistan, Ceylon, Nepal, Isreal, India and other countries. It was brought from Japan in 1959 and introduced in Cuttak (Orissa). The fishes has been transplanted in various states all over the country.

It feeds on phytoplankton and thrives on rice bran and flour. The species naturally breeds in rivers in China and Japan. But in India silver carp was bred in ponds by hypophysation in 1962. (Alikunhi et al. 1963).

Silver carp grows very fast and attained the weight of 2-5 kg. It is cultured in combination with Catla catla in ponds and has been shown faster growth rate than the Indian Carp. The maximum yield was obtained when silver carp and Catla catla were stocked in the ratio of 2:1.

Ctenopharyngodon idella

This fish is commonly known as the grass carp. It is also a food fish. It is known as the "white Amur" in Russia. It is the natural inhabitant of China and Russia. It was brought in India in 1909, from Hongkong and introduced into the ponds of Central Inland fisheries Research Station (CIFR) Cuttack, (Orissa), in 1962. They were bred, successfully by hypophysation and have now been introduced in several status of the country. Male fish matures when two years old and the female when it is three years old. But the off spring obtained as a result of hypophysation in India matured within 2 years in ponds at Cuttack. The individual fish can attain a weight of upto 7kg. when three years old.

Grass carp can be cultivated in fresh water, polluted water and slightly brackish water. The grass carp are suitable for culture in combination with various other carps such as Silver carp, Scale carp, Mirror carp, Rohu, Catla and also with the Tilapia and the Gouramis. This culture is kown as polyculture or composite fish culture.

Osphronemus gourami

It is commonly known as gourami. This fish was brought to Tamilnadu from Java and Mauritius in 1865, but was successfully established in 1916. The fish is introduced into Nilgiri and Kolkata.

This fish cannot survive below 15°C. The fry feed on insect larvae and crustaceans but the adult is a herbivore. Gouramies spawn all the year round in warm waters of the tropical region. The fish naturally breeds in ponds and streams. Male and female take part in nest building. Both parent guards the nest it is a food fish and the flesh has an excellent flavour. It is a popular fish culture in Chennai but has a slow growth rate. It has now been introduced into Bombay Bengal, Orissa and a few other states. Gourami may survive in water with low oxygen content because of their air breathing organs.

Tilapia mossambica

These fishes are originally found along the east coast of Africa, but it has been transplanted to several countries and reared in ponds. In India Tilapia was brought by the Central Marine Fisheries Research Institute, Mandapam from Bangkok in 1952, Tilapia feeds on Zooplankton and phytoplankton but adult is chiefly a herbivore. They are commonly called mouth brooding cichild. They can be cultured in several environments including saline waters. It becomes mature when about two months old growing to a size of 80-90mm. Breeding takes place throughout the year except in sub-tropical climate, where reproduction is interrupted during cold season. The male prepares a shallow nest where female lays the eggs which are then fertilized by male. Female picks up the eggs in her mouth exhibite buccal incubation. Tilapia is a fast growing hardy fish but does not survive at a temperature lower than 10°C. It is not suitable for culture along with carp. When cultured in sewage water mixed with fresh water. It is reported to yield 7000 kg/hec./yr. In Indonesia. It can also be successfully cultured in rice fields. Tilapia also serves well as a forage, fish and when cultured with Channa straiatus in ponds. The yield is considerably increased as it is utilized as food by the murrels. These fishes are not suitable for culture along with carps because it feeds on fry carps. So the production of carps is therefore, adversely affected.

Monosex culture of tilapias by keeping males and females in separate ponds is profitable because they grow in size instead of reproducing at an early age. In such culture large sized specimens upto the length of 300mm and weight of 454 gm. are being obtained by the end of first year.

Puntius javanicus

The common name is Tawes. It is a food fish and was imported from Indonesia in 1972. It is a column feeder and was introduced for weed control. These fishes can feed on the larvae of mosquitoes also. It was introduced for the first time into Kalyani, where it breeds in pond with riverine conditions. Hypophysation has been practised for breeding purpose in India. It is cultivated mainly for control of vegetation in the fish pond and included in polyculture only upon the non-availability of the grass carp. Contrary to grass carp, puntius requires well oxygenated water. As such it is unsuitable for culture in polluted water, where the grass carps are selected for eradication of aquatic weeds. Fish breeds naturally all the year round in the dry bundhs of India. The cultivation of Tawes is not recommended for selling purpose. It has been reported that a single female lays 100000 eggs weighing about 600 g. The hatching period is 14 hrs.

Gambusia affinis

It is commonly called the mosquito fish. It was imported from Italy in

1928 and it is now widely distributed in various parts of India. They feed effectively at the surface of the water on the wriggling larvae of mosquitoes. The fish becomes sexually mature at the age less than a year when it is an inch long only or less.

Labistes reticulates

It is commonly known as guppy. It was brought from South America in 1908. It is a larvivorous fish and becomes sexually mature at the age less then a year. The males may be of variable sizes at the same age. This is well distributed in Rameshwaram. Thanavur, Chennai, Malabar and Southern parts of India (This species has been reported for the first time from Kulbhor river near Lohi in Maharashtra).

Tricogaster

It is commonly a aquarium fish. The common name is three spot gourami. It was imported from Frnace, Indonasia, Malaya and China. It is a larvirorous fish. The fish is about 8 to 10 inch length, is with barbels.

Simies fighter

It is a small fish and originally found in spain. The body is yellow in colour and horizontal bands are present. In breeding season, male becomes dark in colour. It is an omnivore. Fighting of males like cack fighting is a competitive sport and public display.

Petrophyllum scalare

The common name is Angil fish. It is a small fish found in Amazan and Gunia. It is larvivorous black beautiful fish.

CONCLUSION

The exotic fishes are being successfully cultured in different parts of India, without any harm to indigenous forms but some times the balance of natural fauna is upset . Tilapia is incompatible with the major carps with respect to food and space.

Chapter 22

Cultivable Fishes

INTRODUCTION

Fish cultivation in India is expanding first under the continuous pressure of growing population. Traditional method are modernized and the farmers are trained to export vast resources of inland water. There is a large cultivable area of stagnant fresh water in India in the form of ponds, tanks as which only small part is utilized for fish culture. Harvesting cultivable pieces of water under scientific method of fish culture offers a large scope for the development of fishes. According to Jhingran 1976, the country stood on the threshold of aquaplosion. Experimental intensive culture has yielded record production of 9000 kg/he/year. Now fish culture has become an established rural industry of great commercial interest.

The cultivation in India is centered mainly around the carps, which are cultivated more or less extensively. However, there is an urgent need to modify the present image to intensive fish production. Majority of the culturable fresh water carps belong to the order cypriniformes owing to the peculiar digestive system which is stomach less, they feed at shorter intervals than do may other fishes. The abdomen in non-keeled and non-serrated, Air bladder and weberian apparatus is present. Instead of jaw teeth, pharyngeal bones with grinding teeth are present. The body is either naked or covered with cycloid scales.

CHARACTERISTICS OF FISH CULTURE

i) High growth rate.
ii) Good demand.
iii) Easier to handle and harvest.
iv) Reproduction under confined condition.
v) Compatibility with other species under culture.
vi) Ability to live in changing physico-chemical condition such as temperature, pH turbidity, carbon dioxide and dissolved oxygen.
vii) Efficiency to utilize and convert organic production of the water into fish flesh.

TYPES OF CULTURABLE FISH

Indian Major Carps

Catla catla

Catla has deep body with conspicuous head, upturned mouth, non-fringed lips, devoid of barbels and a broad dorsal fin with 14-16 branched rays, are the identifying features. If feed on zooplankton of the pond surface using large gill rakers but young ones feed not only on zooplankton but also on phytoplankton. It is a fast growing species. It breeds naturally in rivers during the rainy season. The artificial propagation by hypophysation is possible. The eggs are non-adhesive and float in the water. It is widely distributed in northern, India, Pakistan, Bangladesh and Burma.

Labeo rohita (Rohu)

This fish is found from West Pakistan across upper India to Burma. It has small and pointed head, small terminal mouth with fringed lower lip, dorsal fin with 12-13 branched fin rays and the dull reddish scales on the body. Finger lings feed on vegetable debris and microscopic plants. Spawning naturally occurs in rivers during monsoon season though hypophysation is also being successfully done. The growth of fish is not as rapid as that of catla. This fish is also suitable for culture in slightly saline water.

Cirrhina mrigal (Mrigal)

The fish *Cirrhina mrigal* is found in the rivers of Uttar Pradesh, Madhya Pradesh, Orissa and South India. It is represented from Pakistan, Bangladesh and Burma also. A linear body, small head with blunt snout, sub-terminal mouth with thin non-fringed lips, dorsal fin with 12-13 branched rays and a bright silvery body having golden tinge are its identifying features. It is a bottom feeder on decaying organic and vegetable debris young feeds on zooplankton. It growth in the first year is about 30 cm. Time of maturity, breeding and distribution are similar to that of catla and rohu.

Minor Carp

Labeo calbasu (Calbasu)

It is commonly known as black rohu. Its body is bluish-green small tapering head with sub-terminal fringe lipped mouth four black barbels and dorsal fin with 12-13 branched rays are its identifying feature. It is an omnivorous bottom feeder on detritus of animals. It is attains a maximum size of 1 m. and growth in the first year is about 30-35 cm. and weight 450 gm. Maturity and breeding habits are similar to Indian major carps. The fish *Labeo calbasu* in found in rivers of India,

Burma, Pakistan, Bangladesh and East Bengal. It is suitable for cultivation in ponds and in slightly saline brackish water.

Labeo bata (Bata)

Labeo bata is mainly found in rivers of North India up to the Godavari. Presence of greenish iridescence at the base of the scales and dorsal fin with 9-10 branched rays are the identifying features of this species. It is an omnivorous bottom feeder on phytoplankton and filamentous algae. Finger lings stocked at the rate of 1 to 1-5 lakh per hectare attain in 25 to 32 cm. in lengths attain maturity in 9-10 months, to produce, ova at a rate of 286 per gm. of body weight of fish.

Cirrhinus Cirrhosa (White Carp)

It has a small head with blunt snout and thin lips. Its dorsal fin in with 14-15 branched rays, the first few rays are much elongated. It has a silvery body and scales with reddish dark except on the abdomen. It is a bottom feeder on detritus and occasionally on zooplankton. The maximum size attained is 45 cm. It is found in abundance in the Godavari, Krishna and the Cauvery.

Labeo fimbriatus (Fringe-Lipped Carp)

It is commonly known as Cauvery carp. Deep body, fringe lips, dorsal fin with 15-18 branched rays and presence of reddish spots in the scales of middle row are its identifying features. It is bottom feeder and feeds occasionally on filamentous algae and zooplankton. Maximum size attained is 90 cm. It is widely distributed in rivers of South India.

Labeo rohita (Pig mouth Carp)

Deep salty colour, prominent snout, sub terminal mouth with fringed lower lip and dorsal fin with 12-13 branched rays are its identifying features. It feeds on detritus, copepods, rotifers, algae and pieces of higher plants. It attains a maximum size of 60 cm. It is widely distributed in rivers of South India especially in the Cauvery.

Puntius sarana (Sarana)

It is an omnivorous feeder on detritus filamentous algae, micro-vegetation, worms, insects and gastropods. Maximum size attained is 30 cm. It is common in East Indian rivers.

Exotic Fishes

Hypophthalmichthys molitrix (Silver Carp)

It is Chinese carp introduced in 1959 by importing from Hong Kong and Japan. An up turned mouth a laterally compressed body with small silvery scales are the identifying features of this species. It is surface feeder, feeding mainly on Phytoplankton. It has a longer digestive canal as compared to grass carp. Young ones of less than 4 cm. feed on zooplankton, maximum size attained is 60 cm. It has a high growth rate, sexual maturity takes place in the second or the third year. Artificial breeding in this species through hypophysation was initially successful.

Ctenopharyngodon idella (Grass Carp)

This species is also a Chinese carp, introduced in India in 1959 by importing from Hong Kong and Japan, Elongated body, broad head with rounded snout, slightly longer upper jaw, moderately sized scales with light greenish tinge and short intestine are its identifying features. Normally it feed on aquatic weeds and terrestrial grass. Newly hatched ones feed on zooplankton including mosquito laeve. Change of carnivorous to herbivorous feeding habit is also observed. It has a very fast growth rate. The maximum growth observed in the first year is 90 cm. and 6.9 kg. Maturity and breeding habits are similar to that of silver carp.

Cyprinus Carpio (Common Carp)

It is a common carp characterized by a deep body and short head. Three varieties are distinguished as

a) *Cyprinus carpio* var *communis* (scale carp) characterized by small scales.
b) *C. carpio*, *Specularis* (Mirror carp) with large shiny and scattered scales.
c) *C. carpio ladies* (Leather carp) by the leathery appearance due to the absence of scales.

C. Carpio var- communis, a Chinese carp was introduced in India in 1939 and 1957 from Sri Lanka and Thailand. White scales carp and mirror carp thrive in plains. Leather carp is mostly confined to cold upland waters. It is omnivorous bottom feeder on larvae of insects, worms, mollusc stalk and leaves of submerged plants and occasionally on zooplankton. Maximum size attained being 75 cm. and 6 kg. It is a fast growing and fairly hardy fish. The eggs are adhesive to submerged vegetation. Common carp together with silver carp and grass carp are known as sanitary fish as they serve as biological controls for maintaining the ecological balance in fish ponds.

Oreochromis mossambicus (Tilapia)

An exotic fish introduced in India from east coast of Africa in 1952. It is characterized by an anterior spinous dorsal fin and posterior soft dorsal fin. A male is identified from a female by its enlarged upper jaw. Maturity occurs even in two months old individuals. It breeds nearly eight times in a year. Female keeps the fertilized eggs guarded in its mouth. It acts as food for murrel in tilapia-cum-murrel culture.

Tinca tinca (Tench or Doctor Fish)

An exotic fish introduced in India from Europe and Eastern Siberia. It is omnivorous feed on diatoms, blue green algae, green filamentous algae, vegetable debris and other bottom fauna. It attains a maximum size of 40 cm.

Cat Fishes

The cat fishes are air breathing or live fishes as they are capable of directly breathing atmospheric air, they can live for a long time without water and can therefore be transferred live and in fresh condition over long distances. The body is without scales, the upper and lower jaws possess two pairs of long barbels is each. The mouth is non-protractile having jaws with teeth. The adipose fin may not be present. Majority of the cat fishes are predatory, feeding on all pond animals including fish fry.

Wallago attu (Fresh Water Shark)

Large mouth extending beyond the eyes, numerous teeth and absence of adipose fin are its identifying features. It is a largest cat fish growing to a maximum size of 2m. It is highly predatory and is not suitable for pond culture.

Mystus aor (Cat Fish)

Presence of adipose fin, maxillary barbell reaching caudal fin and first dorsal fin reaching adipose fin are its identifying features. Maximum size attained is 1.5m. It is common in the rivers and reservoirs of Northern India.

Mystus seenghala (Cat Fish)

Four pairs of barbells, fairly elongated maxillary barbell reaching pelvic fin, longer upper jaw, deeply forked caudal fin of which the upper lobe is longer than the lower and moderately sized adipose fin are the identifying features of this species. Its maximum size is 1m. It is distributed widely in rivers of Northern India.

Pangasius pangasius (Pangash)

Presence of adipose fin, mouth inferior, maxillary barbels reaching the base of the pectoral fin, small and villiform teeth, long and forked caudal fins are its identifying features. It reaches a maximum size of 1.5 m. It is common in India estuaries occasionally in rivers.

Clarias batrachus (Magur)

Absence of adipose fin, elongated first dorsal, long anal and rounded caudal fins are its identify features. It is distinguished from allied species. It viz. C. macrocephalus by the more pointed occipital process which is rounded in C. macrocephalus attains a maximum size of 40 cm. It feed on algae and higher plants of the bottom.

Heteropneustes fossils (Singhi)

Absence of adipose fin and rounded caudal fin are its identifying features. Its feeding habits are similar to that of magur. It reaches a maximum size of 45 cm. It is suitable for culture is swampy areas.

Silonia silonia (Cat Fish)

Small adipose fin slightly ascending mouth maxillary barbels in grooves, mandibular barbels are embedded in skin and deeply forked caudal are its identifying features. It reaches a maximum size of 2m. and 45 kg. It is common in large rivers and estuaries of India.

Murrel

These fishes are also air-breathing and have a good demand. Elongated and cylindrical body, depressed head, large and protractile mouth are its identifying features. Murrels are suitable for culture in irrigation wells and shallow derelict swaps. Though they have protracted breeding season but the peak breeding is associated with pre-monsoon months.

Channa marulius (Giant Snake Head)

Dorsal and anal fins of this species are long and without spines. It reaches a maximum size of 1.2m. It is suitable for culture in ponds along with Tilapia, the young ones of which serve as food source to this species.

Channa stratus (Striped Snake-Head or Common Murrel)

Striped are present on its body. Attempts a maximum size of 90 cm.

Sport Fishes (Trouts)

The body is stream lined and is covered with scales, however, the head is naked. An adipose fin is present in addition to rayed dorsal fin. The barbels are absent, Trouts feed chiefly on insect, mollusc, fish and tadpoles. However, they are cannibalistic at time.

Salmo trutta fario (Brown Trout)

It is river and lake fish of cold northern hemisphere and got introduced to Himalayas river. Its adipose is above the posterior part of anal. Pectorals do not reach the base of the pelvis, caudal fin emarginate, minute scales, deep brown back with sides somewhat lighter. The upper part of the head and body above the lateral line are marked with numerous dark spot. Its fins are dark. It attains a maximum size of 35 cm. and 500 gm.

Salmo gairdner gairdneri (Rainbow Trout)

A North American lake species got introduced to Nilgiris and high ranges of Kerala. It has a small mouth, positioning of adipose above the anal and emarginate caudal fin. Head and dorsal side are steel-blue in colour with reddish band along the sides. Its belly is light coloured. It has numerous dark spots above the lateral line. The dorsal and caudal fins are marked with dark spots. Other fins are pinkish. It attains a maximum size of 40 cm. and 1.3 kg.

Mahsheers

Tortor (Red-finned Mahsheer)

It is a stout fish with a deep body, smell mouth with narrow gapes. It has fleshy lips, maxillary barbels are longer than rostral barbels, deeply notched caudal fin, dorsal side is grayish green, lateral sided are pinkish with greenish gold above and light green below. Belly is silvery. It is an omnivorous feeder on filamentous algae, weeds, insects molluscs and fish. In attains a maximum size of 1.8 cm and 100 kg. It is commonly distributed in the foot hills of the Himalayas, Bihar, West Bengal, Assam and Madhya Pradesh.

Tor putitora (Yellow-Finned Mahsheer)

It is common in the hill stream of Himalayas. Its back is reddish sap green, head is broadly pointed, length of head is greater than the depth of body. The male has enlarged lips with two pairs of barbels of equal length. The pectorals are shorter than the head; caudal is sharply divided. It is an omnivorous feeder on filamentous algae, insect larvae and water plants. It attains a maximum size of 2.7 m.

Tor khudree (Mahsheer)

The depth of body is equal to the length of head. Anterior most part of the body is pointed. It has fleshly lips, deeply forked caudal fin. Dorsal side and areas above lateral line are black and belly is creamy yellowish-white. It attains a maximum weight of 22 kg. Its eggs are demersal, rough to touch and bright lemon-yellow in colour. Feeding habits and distribution are similar to that of tortor.

Acrossocheilus hexagonolepis (Chocolate-Mahsheer)

It is hill stream fish common in eastern Himalayas and Assam. It has thick lips and red eyes which distinguish this from other Mahsheer. Its head is higher than its width. It has sub-terminal mouth, thick lips and large scales. Its caudal is deeply forked with pointed lobes, dorsal surface of body and head are green. It is omnivorous bottom feeder on aquatic weeds insects and mollusc. It attain a maximum size of 99 cm. and 11 kg.

SELECTION AND COMBINATION OF CULTIVABLE FISH RELATION TO WATER RESOURCES

In intensive fish culture, it is essential that fish species selected should be fast growing and good convertors of natural fish food into fish flesh. They should not compete each other for food. The species found efficient for multi-species culture in India are the Indian major carps. These species are tolerant to each other and put to gather for culture, utilize the ecological niches or zones of the pond efficiently and give higher yield. The different species cultured in ponds are Catla, Silver carp, Rohu, Grass carp, Mrigal, common carps and cat fishes.

BREEDING IN CULTIVABLE SPECIES

Breeding of fishes is the first stage in fish culture. Breeding ponds are prepared adjacent to rivers and natural water resources. After the beginning of monsoon, the spawning of carps takes place in the flood regions of the great rivers.

Breeding can be of two types, natural and induced breeding. The natural bundhs are particular types of ponds where riverine conditions or natural water resources conditions are created for the breeding of fishes taken for culture purpose. Low lying area are most suitable for the construction of these bundhs, large quantity of rain water can easily be accommodated in this area. These bundhs have an outlet which can function as an exit for excess of rain water. The shallow are of these specially designed bundhs are used as spawning ground.

Carps breed near confluences of rivers and streams but they are known to breed in bundhs regularly. The eggs are collected within in 12-14 hours after fertilization and are allowed to hatch under protected conditions. The fry are collected with the help of net. Induced breeding has been made practicable, it stimulates breeders to release eggs and sperms finely from the ripe gonads. The spawn collected from riverine and other resources are never pure because in natural condition. It is always mixed with the spawn of other undesirable species. A pure spawn of desired species of fish can be made available by induced breeding only.

PHYSICO-CHEMICAL CHARACTERISTICS OF POND WATER AND FISH CULTURE

A great many of physical and chemical factors influences directly or indirectly the productivity of pond, where as many of these factors are beyond control an appraisal of their condition in pond. Water is always useful in determining the suitability of the pond for fish culture under taking and depending upon its possible productivity the type of fish culture most suiting such a pond. The Physico-chemical nature of pond has much to do with the productivity of pond.

Light

Light is a complex environmental factor and its penetration into pond water can produce diverse ecological effects. The phytoplankton, zooplankton, organic and inorganic particles and turbidity of water, reflect or absorb the light rays, which supplies energy for the process of photosynthesis in plants. Light control several aspects of plant life like structure, shape, physiology, growth and their local distribution etc. Light has far reaching effects on the animal also. It effects a large number of biological, processes like, reproduction, development, growth, locomotion migration and many physiology processes, but light is necessary for photosynthetic activity of the plant providing oxygen and food to the fish and the organisms of water, serving as food for fishes.

Temperature

The range of temperature prevailing in pond water imposes a definite limitation upon the choice of fish to be cultivated, because different fishes have different ranges of temperature-tolerance. Also reproduction, the growth and breathing rates of fishes are directly dependent upon temperature variation.

The different fishes that are suitable for a certain temperature range of water, keeping in view the spawning and growth requirements.

Water temperature 10°C .... Salmon
Water temperature around 10°C ....Pike.
Water temperature around 12°C .... Perch
Water temperature in range of 20-30°C .... Tilapia
Water temperature in range of 20-37°C ....Major carps

The other temperature are not withstood by certain fishes and therefore, these fishes could not be cultivated in such water.

Water temperature 8°C	**Tilapia cannot survive.**
Water temperature around 10°C-12°C or less.	Tropical fishes cannot survive
Water temperature below 16°C	Indian Major Carps cannot survive.
Water temperature above 40°C	Indian Major carps. Chinese grass carp and Silver carps cannot survive.

Turbidity

Turbidity is an important factor. Turbidity, when it is due to plankton growth is a good sign that all does well in the pond water. When it is due to suspension of silt and clay particles, turbidity may be harmful for many fishes. Some fishes are adapted to live in such water and tolerate a certain range of turbidity for example, Indian major carps can tolerate turbidity upto 2000 ppm. In general turbidity affects penetration of sunlight and therefore is unfavourable to photosynthetic activity in the pond water affecting primary productivity.

Dissolved Oxygen Content

Aquatic living organisms need dissolved oxygen in water for respiration and in exchange release carbon dioxide. Dissolved oxygen is also an important factor but different quantities of dissolved oxygen is suitable for different fishes. Photosynthetic plants also provide oxygen to the medium. At 20°C salmonids require 9mg/litre of dissolved oxygen for aquatic breathing whereas carps need 6-7 mg/litre only. Fishes may die due to asphyxia when there is depletion of dissolved oxygen.

Variations in concentration of dissolved oxygen, may occur due to several parameters.

a) Cloudy days reduce it.
b) Higher temperature decrease it.
c) Higher salinity lowers it.
d) Decomposing organic matter.

Sewage fertilizers and manures may also bring about deoxygenation of water. Deficiency of oxygen, especially during late hours of the night, is fatal to the fish. On cloudy days, photosynthesis is reduced and causes, oxygen deficiency at night. Oxygen deficiency in lakes and ponds causes migration attack of parasites fungus as diseases and death due to suffocation.

pH Level

The pond water is graded as follows depending upon its pH level.

pH ....7-Water is neutral
pH .... 7-14 Water is alkaline.
pH ....0.7 Water is acidic

Cultivable fishes tolerate different ranges of pH but sudden fluctuations in pH are harmful. In general natural or slightly alkaline (pH-7.8) Water is more productive than acid water, pH below 4.5 or above 9 is particularly injurious and unproductive. Acid water has following adverse effects.

a) It reduce growth rate by decreasing the appetite of fish.
b) It increase chances of microbial infection, occurs of disease and attack of parasites.

c) It has toxic effects.
d) It lowers dissolved carbon dioxide content reducing photosynthetic activity.
e) It removes lime from water, thereby affecting growth of skeletal tissues and scales.

Carbon dioxide

Carbon dioxide comes into the pond-water through decomposition of organic materials, respiration of various organisms and from the atmosphere. When dissolved in water, carbon dioxide is useful for the green plants in the synthesis of carbon dioxide is useful for the green plants in the synthetic of carbohydrates mainly. Excess is harmful to animals. Carbon dioxide reacts with water forming carbonic acid, which dissociates into H^+ and HCO_3^- ions. In the presence of carbonic acid lime-stone, dissolves forming calcium bicarbonate which influences the organic communities of water.

SBV Factor

(Calcium bicarbonate alkalinity) or Calcium hardness.

Calcium hardness of water is more productive than ordinary water. SBV water is a measure of the bicarbonate of calcium in winter. SBV below 0.1 is rated very poor, between 0.1 to 0.3 poor, between 0.3 to 1.5 averages, above 1 to 5 rich from the point of view of fish cultivation. The bicarbonate of calcium is useful in two ways.

a) It stabilizes the pH of water.
b) It provides a ready reserve of free carbon dioxide for photosynthetic activity.

Nitrogen and Phosphorus

0.2 ppm of dissolved nitrogen is considered favourable for better productivity. The optimum limit of nitrogen can be in the range of 0.3 to 1.3 ppm 60-120 ppm of phosphorus is supposed to be ideal for high productivity.

Aquatic Fauna and Flora

Pond should be rich in aquatic fauna and flora. It definitely increase the productivity or the pond phytoplankton and zooplankton that are consumed by the fishes. The submerged plants are also as great use. The bottom fauna and the bottom debris also serve as food for many fishes.

Depth of Pond

The depth of water is an important factor. Penetration of light and temperature of water depend on the depth of water. Photosynthetic activity and several physico-chemical conditions of pond are affected. Sustaining of required amount of biota depend on the depth of the pond. The light is modified by water medium in respect to intensity and spectral composition. Different rays penetrate upto a definite depth of water. Distribution of phytoplankton and zooplankton also depend on the depth of the pond.

Water Movements

Water is a per-requisite for breeding of fresh water carps, though the fishes which can thrive well in stagnant ponds.

Movement of water due to wave action, currents and drainage system favours the productivity of pond. Fast flowing waters however, have negative attributes because they bring about erosion of soil and turbidity and take away the fish food necessary for the fish production.

Alkalinity

The water becomes alkaline due to the presence of calcium magnesium, sodium, potassium etc. Alkalinity is the capacity of water to react with an acid. There are various grades of alkalinity of water which changes with rain, maximum of clay in water, change in photosynthetic activity of aquatic plants etc. The level of alkalinity is a measure of productivity of water.

IMPROVEMENT MANAGEMENT OF POND FOR FISH CULTURE

The productivity of the pond depends upon soil base and can greatly enhanced by

Cleaning the Pond Bottom

The pond should be cleaned at the suitable intervals.

Dredging the Pond

Old pond is neglected and dredging is necessary to restore is productivity. In such a pond the accumulation of the mud on the bottom is as great as it cannot be ordinarily removed. If care is not taken in proper time, such a pond soon turns into a swamp which is not only unfit for fish culture but is also too expensive for any restoration.

Increasing the Pond Depth

Drying of the pond if often combined with working of the bottom so as to deepen it. This is necessary to remove the deposition of mud at the bottom, which tends to make the pond shallower in course of time. This may destroy the useful true pond mud as well by bringing the deep sterile layers of soil to the top of the bottom.

Control of the Vegetation

All undesirable plants and weeds must be removed regularly by using various methods of weed control. The floating weeds and the submerged weeds does harm in several ways the pond productivity.

a) It provides shelter for food competitors for fish and their enemies.
b) It increase silting, restricting fish movement.
c) It shades water, cutting the sunlight penetration.
d) It stop water from warning impending oxygen circulation.

It turns the Pond into a Swamp

The elimination of the excessive weed is necessary to improve the fish production of the pond.

Control of weeds is an important part of pond management. It involves removal of excessive weds to a level, when they become not only harmless but useful to fish cultivation. It must not be forgotten the moderate growth of aquatic plants is not only useful but necessary for pond productivity.

Use of Lime

The first step in fertilization of a pond is the application of lime. Lime is used in many forms such as lime stone, quick lime, calcium cyanamide, caustic lime etc. Quick lime is very effective, but has a toxic effect upon fish. Calcium cyanamide is more toxic than any other form. Liming is required under any of the following conditions.

a) When pH level is very low.
b) When SBV alkalinity is very low.
c) When decomposing organic matter is abundant.
d) When there is a threat of dissolved 02 depletion.
e) When prophylactic measure are needed.
f) When there is an outbreak of a disease is threatened or there are fish enemies.
g) When bottom soil is poor.

For **a** and **b** lime stone is useful for **e** and **f** quick lime, caustic lime or calcium cyanamide are more useful.

There are two ways of liming. In the one the bottom of the pond after is has been completely drained and dried is treated with lime, to destroy parasites, disease causing agents and fish enemies as well as for general disinfection. This is also helpful for improving the soil. The other method is to lime the pond water in order to nullify the acidity of the pond water.

The principle of action underlying liming is to increase the bicarbonate contents of the water so the buffer conditions prevail. Use of lime should however, be done with caution. Excess of lime is troublesome because of its toxic effect.

Lime is generally used and is spread on the pond bottom for 10-15 days before stocking the pond with fish. It is better to leave the pond dry for at least two weeks after liming. The liming is cheap but the advantages are many.

a) Lime is anti-parasitic, destroying parasites living in the water or in the intermediate hosts.
b) It destroys the algae and submerged weeds.
c) It kills the fish enemies such as insect and their larvae.
d) It corrects the acid water to make it slightly alkaline (pH 7-8) thereby making is good for the health of fish on the one hand and for the biological productivity of the pond in general on the other hand.
e) It causes excessive organic suspension to precipitate that are needed by animals and plants respectively.
f) Besides, ca^{+} ion neutralizers the harmful effect Mg, N and K ions.
g) It controls quickly outbreaks of diseases.
h) It has all the useful effects of a fertilizers in addition to make the pond more fertile.

Controlled growth of aquatic vegetation is beneficial for productivity of the fish pond. Many fishes feed on planktons throughout their life, other feed on submerged vegetation and require weeds for depositing eggs in the breeding season. In intensive sunlight the egg of fishes get shelter among the weeds and plants and similarly the larvae and insect etc. Which constitute the food of fish. Vegetation produce oxygen, required for respiration of fishes and other aquatic animals and bring back various nutrients of soil into the food chain. Thus the vegetation in pond play an important role in maintaining the ecophysiology of the pond.

Use of Fertilizers

The fertilizers are used to increase fish production in ponds. The pond fertilization improves fertility in the following way

i) It increases the abundance of natural food in the pond.
ii) It does not cause dietary diseases.
iii) It improves the hygienic condition of the pond by speeding up the decomposition of food residues and excreta.

The fertilizers are used to make the pond fertile and to improve fish production. The fertilizers are classified into two categories.

a) Inorganic fertilizers (Mineral fertilizers)
b) Organic fertilizers (Manures of plant and animal origin).

The main inorganic fertilizers include lime, phosphate, potash, nitrogenous compounds and magnesium fertilizers.

The organic fertilizers comprises green manure, the compost and the manures of the animal origin such as liquid from cattle farms, sewage, dung etc.

For best results, the pond to be fertilized, should be prepared beforehand as follows

i) The acidity of the pond should have been corrected so as to have a neutral or slightly alkaline conditions.
ii) All emergent weeds should have been destroyed and removed.
iii) The floating and submerged weeds should have been properly controlled so that sunlight penetration is not interfered with.
iv) The pond bottom should have a good mud but not a mud of cellulose.

Inorganic Fertilizers

The suitability of any fertilizers is judged from the following criteria.

a) How much increase in the production of plankton it can bring about.
b) How cheap it is?
c) How easily it can be distributed over the pond?
d) How much harmless it is to the fish under cultivation.
e) What after effects it will have on the conditions of the pond.

In order to work out the suitability of fertilizers, tests are necessary. Schaperclaus has suggested a scheme of comparative tests for determination of the merit of a fertilizers.

The five ponds of same size, same stocking and under identical conditions are selected and treated as follows. First is kept as a control and therefore left untreated. Second pond is given a dose of phosphate fertilizer at the rate of 40 kg/hec. likewise the third, fourth and fifth are treated besides the above dose of the phosphate, an additional dose of potash, (50 kg/hec.), lime (500 to 1000 kg/hec.) and nitrogenous fertilizer (50 kg/hec.) respectively. The result will indicate the suitability of the fertilizers.

Organic Fertilizers

The organic fertilizers used show additional advantages such as,

a) Any single manure compensates for practically all the necessary nutrients required in the biological production.
b) It improves the bottom mud.
c) It encourages bacterial growth, in turn favours better production of the zooplankton.
d) It increases the effectiveness of many inorganic fertilizers by providing the necessary organic matter base.

However, there are some points which call for the use organic manures with great caution. The use of organic manure is made in small quantities. The distribution is done in several spread. As for as possible heaping is prevented by uniform spread. The fry and the young fish are the stages which profit the most.

Use of Artificial Food and Feeding

This is one of the principal means of increasing fish yields from ponds, whether the artificial food issued as the exclusive food or supplementary food. Suitability of an artificial food in fish culture depends upon a number of qualities such as.

a) It should be readily acceptable to fish.
b) It has a high conversion rate, that is to say less food is needed to produce on kg of fish flesh
c) It should be cheap and economical. There is a long list artificial foods that have been tried for various species and found to be useful for enhancing the fish production in cultivation. These may be classified in several ways such as food of plant or animal origin, unprocessed food, ordinary foods or day concentrates etc. Success of good artificial food in turn depends upon.
    i) The amount of food given.
    ii) The manner of distribution.
    iii) The time of distribution.
    iv) The density of fish.
    v) Age group of the individuals.
    vi) State of health of individuals.
    vii) Physical condition of water.

Control of Fish Enemies

Various animals including fish are harmful to fish cultivation in so far as they either live upon the cultivated fish species or their eggs and fry or even compete them for food to this category of fish enemies be long all groups of vertebrates as well as insects. Depending upon the population of any of these, a control becomes necessary as part of pond productivity or management.

1. To control frogs and their tadpoles, snakes, birds etc. suitable measures have to be followed. By applying quick lime the eggs frogs can be killed. However, for avoiding the entry of adult frogs and snakes into fish ponds, wire tropes, nets or drag nets may be employed.
2. Birds like herons and cranes are invariably controlled by shooting scaring or netting.
3. Periodical drainage of ponds and drying as pond bottom destroy insect and their larvae that are harmful to fish eggs and fry.
4. Use of toxic substances is quite effective in certain cases.

Control of Disease

The fish can be subject to many diseases particularly when the farming is intensive many fungal, bacterial, viral, parasitic, environmental, dietary and other types of diseases have been encountered in the fish crop. Pond water must be of good quality in order to avoid dangers from lack of oxygen and pollution. Infected ponds and the material used must be disinfected. Sometimes the diseases many take the form of epidemics and large scale mortality may be caused ponds should be hygienically maintained. Frequent use of benzalkonium chloride and potassium permanganate disinfects the pond. Use of lime if also good for this purpose. Fishes should always be supplied with rich food.

Skimming Intermediate Fishing throughout a Year

Intermediate fishing increase total fish production in a year. By intermediate fishing operations the population of fishes in the pond should be reduced. It provides better scope for individual growth. When a fish attains marketable size it should be removed. Successive multiple production in one year should also be attempted, if climatic condition favours.

By Calculating the Correct Stocking Rate

Taking other conditions favourable such as suitable choice of fish species and a well maintained pond with fertilizer and artificial food, with fish production depends upon the number of weight of the fish with which the pond is originally stocked.

The stocking rate is therefore yet another very important means of increasing production of fish.

The following considerations from the basis for the stocking rate.

a) High stocking density results in increased total production as there is better utilization of the available food. But in such case the weight and size of individual in the harvest in reduced, that is the disadvantage of the market value.

b) Low stocking density gives rise to a crop with a considerable increase in the market value, since the individuals produced are large and with increased weight.

c) Stocking with small sized fish leads to higher yield per unit area than with large sized fish.

Correct stocking rate for a pond is that optimum quantity of a fish which when stocked in the pond results in a given time into a production which is higher in quantity and quality of fish under maximum economy.

Economy

The Stocking rate is determined keeping in view.

i) The size of pond.
ii) The productivity of pond.
iii) The size and kind of species of fish selected for cultivation.
iv) The time period for raising the culture of crop.
v) The nature of crop desired.
vi) The manner of raising the culture extensive, intensive or semi intensive.

It must be remembered that there will be more production quantitatively when young fish are raised than when old fish are raised in the same pond.

Control of Physico-Chemical Factors

All the physico-chemical characteristic of water should be thoroughly checked and corrected during the entire period of cultivation. This is important because their concentration may change depending upon the environmental conditions. To cite a few example, the pH of pond water may changes in rainy season owing to flood water binging in the acidity, oxygen concentration may fall below the required level in cloudy days and should be corrected quickly by aeration of by other means.

Control of Weed Fishes and Predators

Weed fishes consume the aquatic vegetation and thus compete with the fish water under cultivation. They appear in the fish under cultivation. They appear in the fish and during the rainy season and should be removed by either, repeated netting or by draining the pond. They may also the killed by using mild poison.

Predators like aquatic insect the carnivorous fishes, should be prevented from entering the pond or be removed, before stocking the pond. They may be controlled by using natural poison or by chemicals.

References

Alabaster, J.S. and Lloyd, R. 1980. Water quality criterion for fresh water fish. Butterworths, London, Boston, 297-298.

Alikunhi, K.H. 1956. Fish culture techniques in India. In progress of fisheries Development in India. CIFRI and Dept. fish Orissa 63: 73.

Alikunhi, K.H. 1957. Fish culture in India, Fm. Bull. India. Coun. Agric. Res. 20: 144.

Alikunhi, K.H., Datta, A., Singh, V.D. and Dubey, G.P. 1964. Observations on the breeding of carps in bundhs, near Nowgong, Madhya Pradesh during July. Aug. Bull. Cent. Inst. Fish education. Bombay, 1: 22.

Allen, K.R. 1951. A study of a trout population Fish Bull. N.Z. 10: 1-238.

Arnold, E. 1966. Marking fish with dyes and other chemicals Bureau of sport fisheries and Wild life. Technical paper No. 10-44 p.

Bagenal, T.B. 1963. Variations in plaice fecundity in the Clyde area, J. mar. biol. Ass. U.K. 43: 391-399.

Bagenal, T.B. 1966. The ecological and geography aspects of the fecundity of the plaice, J. mar. biol. Ass. U.K. 46: 161-186.

Bardach, J.E., Ryther, J.E. and Molaoney, W.O. 1972. Aquaculture The farming and Husbandry of fresh water and Marine organisms, New York, John Wiley & Sons.

Bell, F.W. and Centerbery, E.R. 1976. Aquaculture for the Developing Countries, Camb. Ballin. Publis. Co.

Benerjee, P., Bhattacharya, P.S. and Nath, P. 1989. Purification and properties of pituitary gonado trophic hormone from Indian teleosts. Fresh water murrel channa punctatus and carp. Catla, Gen. Comp. Endocr, 73: 118-128.

Benerjee, R.K., Saha, S.k., Sen, P.R. and Srinivasan, K.V. 1989. Ecological studies on town refuse in the context of environmental pollution abatement and fish production. J. Inland Fish. Soc. India. 21(2): 25-30.

Bhargava, S. 2000. Sewage Treatment Refer. Cour. Zool. Sagar, p. 26-38.

Biswas, K.P. 1996. A text book of Fish Fisheries and Technology 2nd Edi- Naresh Publis Hous. Delhi (Book).

Blackburn, R.D. 1968. Weed control in fish ponds in the United States, FAO, Fish Rep. 44(5): 1-17.

Bose, P.C. 1944. Calcutta Sewage and Fish Culture. Proc. Nat. Inst. Sci. India, 10: 443-450.

Boyd, C.E. 1970. Vascular aquatic plants for mineral nutrients removed from polluted. Water. Econ. Bot. 23(1): 95-103.

Chaudhury, H. and Tripathi, S.D. 1976. Problems of warm water fish seed production. Advance in Aquaculture. Tech. Conf. Aqua. Kyoto. Japan. 127-134.

Chua, Thia. Eng. Elsie. Tech. 1990. Aquaculture in Asia. Asian Fisheries Society. Ind. Bran. 13-30.

Das, K.M. and Tripathi, S.D. 1979. Feed Protein Utilization and growth of

Ctenopharyngodon idella. Ind. Sci. Cong. 66(3): 85.

Das, I. Sita and Ray, A.K. 1989. Growth performance of the Indian major carp Labeo rohita on duck weed. Incorporated pelleted feed, a preliminary study. J. Inland Fish Soc. India. 21(1): 1-6.

Davis, C.W. 1971. Marketing small Atlantic herring with biological stains. Prog. Fish culturists. Vol. 33(3): 160-162.

David, A.N., Rao, G.S. and M.F. Rahman, 1970. J. Inland Fish Soc. India. 2: 159-160.

Dwivedi, S.N. 1980. Fish hatchery and fish breeding unit for raising fish seed in controlled environment in India. Today & Tomorrow. 8(4): 167.

Edwards, P. 1980. Food potential of aquatic macrophytes ICLARM stud. and reviews Manila. 5-51.

Ganapathi, S.V. 1972. Re use of waste water. J. Sci and Ind. Res. 31: 390-393.

Gaur, A.C., Neelakantam, S. and Dargan, K.S. 1984. Organic manure, ICAR Publication, New Delhi.

Gerking, D.S. 1978. Ecology of fresh water fish production. Black well Scientific Publication. 17: 424-446.

Ghosh, A., Benerjee, M.K. and Hanumantha Rao 1973. Some observation on the cultural prospects of Silver carp in sewage feed ponds. J. Ind. Fish. Soc. India. 5: 131-133.

Gopal Krishan, V. 1961. Observation on infection dropsy of Indian major carps and its experiment induction. J. Scient. Ind. Res. 20: 357-358.

Green Wood, P.H., Rosen, D.E., Weitzman, S.H. and Myers, G.S. 1966. Studies of teleostean fishes with a provisional classification of living forms. Bull Amer. Mus. Nat. Hist. 131(4): 1-455.

Gupta, S.P. 1961. A reference list of trematodes. Parasites of fresh water fishes of India with a discussion on their systematic position. Ind. J. Helminth. 13(1): 35-60.

Gupta, O.P. 1979. Aquatic weeds, their menace and control. Today and Tommarrows Publishers and Printers, New Delhi.

Hajra, A. and Tripathi, S.D. 1985. Nutritive value of aquatic weed (Spirodela polyrhizalinu) in grass carp. Ind. J. Anim. Sci. 55(8): 702-705.

Hajra, A. 1987. Bio-chemical investigations on the protein calorie availability in grass carp, from an aquatic weed in the tropics. Aquaculture, 61: 113-120.

Halver, J.E. 1976. The nutritional requirements of cultivated warm water and cold water species FAO. Technical conference on Aquaculture. Kyoto Japan, FIR/Conf. 76/F. 131.

Hora, S.L. 1921. Indian cyprinoid fishes belonging to the genus. Garra. Rec. ind. Mus. 22: 635-643.

Hora, S.L. 1951. Notes on the fishes of Indian Museum. XL. VII. Red. Indian. Mus. XXIV.

Hora, S.L. and Pillay, T.V.R. 1962. Hand book on fish culture in the Indo Pacific region. FAO Fish Tech. Vol. 14, FAO.

Hora, S.L. 1923. Observations on the fauna of certain torrential streams in Khasihills, Rec. Ind. Mus. 25: 579-598.

Hora, S.L. and Pillay, T.V.R. 1962. Hand book on fish culture in the Indopacific fisheries region. FAO, Biol. Tech. Pap. (14). 203.

Huet, M. 1972. Text book of fish culture breeding and cultivation of fish. England. Fishing News Book.

Iyenger, H.D.R. 1962. Studies on Paddy-cum fish culture at Hesserghatta fish farm in Mysore State. J. Bombay. Nat. Hist. Soc. 59(1): 301-305.

Jhingran, A.G. and Ghosh, A. 1988. Aquaculture as a potential system of sewage disposal. J. Inland Fish Soc. India. 20(2): 1-8.

Jhingran, V.G. and Sehgal, K.L. 1978. Cold water fisheries of India. "Care of brood stock" Inland fisheries society of India. 109-111.

Job, T.J. 1940. Note on the geographical distribution of the larvicidal propensities of *Horaichthys setnai* Kulkarni. J. Bombay nat. Hist. Soc. 42: 201-202.

Job, T.J. 1943. Mosquito-larval eating fish. Assam. Review. 111-115.

Katterns, L.B. 1971. Tropical fish and Aquaria, London (Book).

Kapur, K. 1981. Spawning of major Indian carps a review. Actahydrochim, Hydrobiol. 9(2): 163-182.

Khan, H. 1942. Spawning of carps and their spawning grounds. J. Bombay. Nat. Hist. Soc. 43(3): 416-427.

Khanna, S.S. 1996. An introduction to fishes central book depot Allahabad. 180-183.

Kumar, D.J., Farkas, J. and Sinha, V.R.P. 1986. Bacteria from diseased fresh water fish from India. Aquaculture. V. 113-118.

Kumar, D. and Dey, R.K. 1988. Bacterial septicemia in Silver Carp (Hypophthalmichthys Molitrix) Vet. Arhiv. 58(6): 277-283.

Kutty, M.N. 1979. Utilisation of Sewage for fish culture. Inte. Sympo. Tropi. Eco. Kuawlampur, Malaysia. 106.

Kutty, M.N., Natrajan, V. and Venkataramanujam, K. 1979. Recycling sewage for fish culture. Proc. Symp. Inland Aquaculture. CIFRI Barrach Pune. 99-100.

Kyle, H. 1993. The biology of fishes. Himalaya, Pub. New Delhi.

Lagler, F.K. 1956. Fresh water fishery biology. W.M.C. Brown Company, Dubuque Towa.

Lal, M.B., Bhatnagar, A.N. and Uniyal, J.P. 1966. Adhesive modification of a hill stream fish. Glyptothorax pectinopterus. Proc. Nat. Acad. Sci. 36E. 109-116.

Lawrence, J.M. 1968. Aquatic weed control in fish ponds FAO. Fish Rep. 44(5): 76-91.

Lisifa. 1990. Recent Advances in fresh water Aquaculture in China Asian. Fisheries Society Indian Branch. 141-161.

Ludwig, G.M. and Lange, E.L. 1975. The relationship of length, age and age length inter action to the fecundity of the northern mottled sculpin cottus. As bairdi, Trans Am. Fish. Soc. 104: 64-67.

Mackenzie, R.A. and Hall, W.T.K. 1976. Dermal Ulceration of mullet (Mugil cephalus) Aust Vet. J. 52: 230-231.

Mammen, T.A. 1962. Live fish transport using modified splash less carrier with compressed in aeration. Training course on live fish transport. Govt. of India.

Mathur, B.B.L. 1952. Notes on fishes from Rajasthan, India, Rec. India Mus. 105-110.

Menon, M.A.S. 1952. On a collection of fish from Manipur, Assam Rec. Indian Mus. XLVII, 225-237.

Mohanty, S.N., Swamy, D.N. and Tripathi, S.D. 1990. Growth nutritional indices and carcass composition of the Indian major carp fry. *Catla catla*, *Labeo rohita* and *Cirrhinus mrigal* fed four dietary protein levels.

Hungarico. 6: 211-217.

Nair, R.V. 1953. Key to the identification of the common clupeid fishes of India. J. Zool. Soc. India. 108-135.

Natarajan, V., Venkataramanujam, K., Kutty, M.N. and Rajesekharan. 1977. Fish Production through recycling waste waters, seminar 16. 1-16. 10.

Pandian, T.J. and Muthukrishnan. 1990. Methods for Gene and chromo. Some manipulation in fish. A training programme Govt. of India (Madurai). 1-58.

Pandey, A.K. and Sandhu, G.S. 1992. Encyclopedia of fishes and fisheries of India. Vol. III, IV.

Patra, B.C. and Ray, A. 1988. Observation on the influence of three organic manures on the growth and yield of Indian major carps under pond culture. J. Inland Fish Soc. India. 20(2): 61-63.

Patnaik, S., Das, K.M. and Pani, K.C. 1988. Large scale clearance of Water Lily (Nymphaea nauchali) by 2,4-D and subsequent management of the ponds, through fish culture. J. Inland fish Soc. India. 20(1): 57-62.

Pillai, T.V.R. 1968. Warm water pond fish culture FAO. Fisheries Rep. No. 44, Vol. 5.

Pullin, R.S.V. and Shehadan, Z.H. 1980. Integrated Agriculture Aquaculture farming systems ICLARM Conf. Proc. Vol. 4.

Philipose, M.T. 1968. Present trends in the control of weeds in fish culture waters of Asia and the Far. East. FAO. Fish Rep. 44 (5): 26-52.

Philipose, M.T. 1972. Fifty years of Aquatic weed control in India. The Hague. 215-223.

Pruginin, Y. 1968. Weed control in fish ponds in the Far East, FAO, Fish, Rep. (44) 5: 18-25.

Raiti, D.F.S. and Hall, W.B. 1967. On fecundity of the red fish *Sebastes marinus*. J. Cons. Perm. Int. Explor. Mer. 31: 237-245.

Ramachandran, V. and Ramaprabhu, T. 1968. Investigation on aquatic weed control with special reference to the use of chemicals. FAO Fish Rep. 44(5): 92-108.

Ranga, M.M. and Shammi, Q.A. 20001. Fish Biotechnology. Agrobios (India), Jodhpur.

Rath, R.K. 1993. Fresh water Aquaculture. Scient. Pub. Jodhpur.

Rounsefell, G.A. 1957. Fecundity of the North America Salmonidae, U.S. Fish and Wildlife Serv. Fish. Bull. 57: 451-468.

Roy, D.N. 1933. On the control of Malaria Mosquitoes in Bengal by the use of predaceous fishes and on the habit of two of them, J. Mal. Inst. India. 1(4): 405-416.

Saha, K.C., Sen, D.P., Roy, A.K. Chowdhury and Chakravarty, S.K. 1957. Studies on the factors influencing spawning of Indian major carps in bundh fisheries.

Sastry, K.V. 2000. Transgenic fish production. Refer. Cours. 2001. PP. 59-6t9.

Sankaran. 1976. Aquatic weeds in South East Asia. 323-327.

Sar, U.K. 1991. A new tool, Simyra, Conspersa for biological control of Nelumbo nucifera in fish ponds. Proc. Nat. Sym. Fresh water Aqua. 117-119.

Sedwick, S.D. 1976. Trout farming hand book, London, Seeley service.

Sharma, U. and Grover, S.P. 1982. An introduction to Indian fisheries. B. Singh Mahendra Singh Publication. Dehradun.

Sharma, B.K. 1990. Integrated Fish farming various possibilities and economic considerations. Techniques of Aquaculture Management in large water bodies. CIFA, Training Report. 66-76.

Shukla, G.S., Upadhyay, V.B. 1998. Economic Zoology, Rastogi Pub. Meerut.

Sheela, S.G., Cheh, J.D., Mathavan, S. and Pandian, T.J. 1998. Construction electrophoretic transfer and expression of ZphyPGH and ZpBrTGH in Zebra fish. J. Bio. Sci. 23(5): 565-576.

Sinha, V.R.P., Jhingran, V.G. and Ganapati, S.V. 1974. A review on spawning of the Indian major carps, Arch. Hydrobiol. 73(4): 578-556.

Sinha, V.R.P. and Gupta, M.V. 1975. On the growth of grass carp in composite fish culture at Kalyani West Bengal Aqua. 24(4): 283-290.

Singh, B.R. 1993. Advances in fish research. Vol. I Naren Publi. Hou. Delhi.

Srinivasan, R. and Chako, P.I. 1952. The control of Aquatic vegetation ponds in Madras with an account of the chemistry water and Soil. FAO Fish Rep. 44(3): 179-197.

Srivastava, C.B.L. 1999. A text book of fishery science and Indian fisheries. II Ed. Kitab Mahal Publication, Allahabad.

Swingle, H.S. 1957. Weed Control FAO Fish Rep. 10: 11-17.

Tal, S. 1974. Preliminary observations on new methods of fish culture in Israel Vol. 26(3): 51-56.

Tan, Y.T. 1970. Composition and nutritive value of some grasses, plants and aquatic weed tested as diets, J. Fish Biol. 2: 253-267.

Tripathi, S.D. and Khan, H.A. 1990. Fresh water Aquaculture in India Asian Fish Soci. Bran Indian. 191-222.

Verma, R.S., Singh, A.K. and Tripathi, N.C. 1991. Irrigating trees with raw sewage water Indian Farm Vol. 41(1): 27-28.

Wain Wright, N. 1969. Cold water Aquariums, London, Frederick Warne Co. Ltd.

Watanabe, T. 1987. Lipid nutrition in fish comparative Bio-Chemistry and Physiology, 73-B, 3-15.

Welcomme, R.L. 1967. Th[illegible]ationship between fecundity and fertility in the mouth breeding cichli[illegible]sh. J. Zool Lond. 151: 453-468.

Woyaha Rovich, E. 1975. Elementary Guide to fish culture in Nepal FAO Publ.

Woynarovich, E. and Horvath, L. 1980. The artificial propagation of warm water fin fishes. FAO, Fish Tech. 20-183.

Yadav, B.N. 1997. Fish and Fisheries Daya Publishing House, Delhi.

Zawisza, J. and Backiel, T. 1970. Gonad development Fecundity and egg survival in *Coregonus albula* Uni-winn Canada. 363-397.